Calculus

Mehdi Rahmani-Andebili

Calculus

Practice Problems, Methods, and Solutions

 Springer

Mehdi Rahmani-Andebili
State University of New York
Buffalo State, NY, USA

ISBN 978-3-030-64982-1   ISBN 978-3-030-64980-7   (eBook)
https://doi.org/10.1007/978-3-030-64980-7

This Springer imprint is published by the registered company Springer Nature Switzerland AGThe registered company address is: Gewerbestrasse 11, 6330 Cham, Switzerland

Preface

Calculus is one of the most important courses of many majors, including engineering and science and even some of the non-engineering majors like economics and business, which is taught in the first semester at universities and colleges all over the world. Moreover, in many universities and colleges, precalculus is a mandatory course for the under-prepared students as the prerequisite course of calculus.

Unfortunately, some students do not have a solid background and knowledge in math and calculus when they begin their education in universities or colleges. This issue prevents them from learning other important courses like physics and calculus-based courses. Sometimes, the problem is so escalated that they give up and leave university or college. Based on my real professorship experience, students do not have a serious issue to comprehend physics and engineering courses. In fact, it is the lack of enough knowledge of calculus that hinders their understanding of calculus-based courses. Therefore, this textbook along with the other one (*Precalculus: Practice Problems, Methods, and Solutions, Springer, 2020*) have been prepared to help students succeed in their major.

The textbooks include basic and advanced problems of calculus with very detailed problem solutions. They can be used as a practicing textbooks by students and as a supplementary teaching source by instructors. Since the problems have very detailed solutions, the textbooks are useful for under-prepared students. In addition, the textbooks are beneficial for knowledge-able students because they include advanced problems.

In the preparation of problem solutions, effort has been made to use typical methods to present the textbooks as an instructor-recommended one. By considering this key point, both textbooks are in the direction of instructors' lectures, and the instructors will not see any untaught and unusual problem solutions in their students' answer sheets.

To help students study the textbooks in the most efficient way, the problems have been categorized in nine different levels. In this regard, for each problem, a difficulty level (easy, normal, or hard) and a calculation amount (small, normal, or large) have been assigned. Moreover, in each chapter, problems have been ordered from the easiest one with the smallest amount of calculations to the most difficult one with the largest amount of calculations. Therefore, students are advised to study the textbooks from the easiest problem and continue practicing until they reach the normal and then the hardest one. On the other hand, this classification can help instructors choose their desirable problems to conduct a quiz or a test. Moreover, the classification of computation amount can help students manage their time during future exams and instructors assign appropriate problems based on the exam duration.

Buffalo, NY, USA   Mehdi Rahmani-Andebili

Contents

Problems: Trigonometric Equations and Identities

Abstract

In this chapter, the basic and advanced problems of trigonometric equations and trigonometric identities are presented. The subjects include trigonometric equations, trigonometric identities, domain, range, period, sine and cosine identities, tangent and cotangent identities, half angle formulas, reciprocal identities, Pythagorean identities, sum and difference to product formulas, product to sum formulas, even and odd formulas, periodic formulas, sum to product formulas, double angle formulas, degrees to radians formulas, cofunction formulas, unit circle, inverse trigonometric functions, inverse properties, alternate notation, and domain and range of inverse trigonometric functions. To help students study the chapter in the most efficient way, the problems are categorized based on their difficulty levels (easy, normal, and hard) and calculation amounts (small, normal, and large). Moreover, the problems are ordered from the easiest problem with the smallest computations to the most difficult problems with the largest calculations.

1.1. Calculate the value of $\tan(2\theta)$ if $\cot(\theta) = 5$ [1].

Difficulty level ● Easy ○ Normal ○ Hard
Calculation amount ● Small ○ Normal ○ Large

1) $\frac{5}{12}$
2) $\frac{5}{13}$
3) $-\frac{5}{12}$
4) $-\frac{5}{13}$

1.2. Determine the value of $\tan(-2100°)$.

Difficulty level ● Easy ○ Normal ○ Hard
Calculation amount ● Small ○ Normal ○ Large

1) $\sqrt{3}$
2) $\frac{\sqrt{3}}{3}$
3) $-\sqrt{3}$
4) $-\frac{\sqrt{3}}{3}$

1.3. Simplify and calculate the final value of the following term.

$$\frac{1 + \cos(40°)}{\sin(40°)}$$

Difficulty level ● Easy ○ Normal ○ Hard
Calculation amount ● Small ○ Normal ○ Large

M. Rahmani-Andebili, *Calculus*, https://doi.org/10.1007/978-3-030-64980-7_1

1) $\sin(20°)$
2) $\cos(20°)$
3) $\tan(20°)$
4) $\cot(20°)$

1.4. Determine the range of m if $\sin(\alpha) = \frac{3m-1}{4}$ and $\frac{\pi}{6} \le \alpha \le \frac{2\pi}{3}$.

Difficulty level   ● Easy   ○ Normal   ○ Hard
Calculation amount   ● Small   ○ Normal   ○ Large

1) $\left[1, \frac{2\sqrt{3}-1}{3}\right]$
2) $\left[1, \frac{2\sqrt{3}+1}{3}\right]$
3) $[1, 2]$
4) $\left[1, \frac{5}{3}\right]$

1.5. Determine the range of m if $\cos(\alpha) = \frac{2m-1}{6}$ and $-\frac{\pi}{3} \le \alpha \le \frac{\pi}{3}$.

Difficulty level   ● Easy   ○ Normal   ○ Hard
Calculation amount   ● Small   ○ Normal   ○ Large

1) $\left[2, \frac{7}{2}\right]$
2) $\left[\frac{3}{2}, \frac{7}{2}\right]$
3) $\left[2, \frac{5}{2}\right]$
4) $\left[\frac{3}{2}, \frac{5}{2}\right]$

1.6. What is the main period of $\cos^2(x) - 5\cos\left(\frac{2x}{3}\right)$?

Difficulty level   ● Easy   ○ Normal   ○ Hard
Calculation amount   ● Small   ○ Normal   ○ Large

1) π
2) 2π
3) 3π
4) 4π

1.7. What is the main period of $\sin^4\left(\frac{3x}{5}\right) + \cos^3\left(\frac{2x}{3}\right)$?

Difficulty level   ● Easy   ○ Normal   ○ Hard
Calculation amount   ● Small   ○ Normal   ○ Large

1) 3π
2) 5π
3) 15π
4) 30π

1.8. Determine the main period of $\sin^4\left(\frac{\pi x}{3}\right) + \cos(\pi x) + 5$.

Difficulty level   ● Easy   ○ Normal   ○ Hard
Calculation amount   ● Small   ○ Normal   ○ Large

1) 1
2) 2
3) 3
4) 6

1.9. Figure 1.1 illustrates a part of the function of $y = \sin(kx)$. Determine the value of k.

Difficulty level   ● Easy   ○ Normal   ○ Hard
Calculation amount   ● Small   ○ Normal   ○ Large

1) $\frac{2}{3}$
2) $\frac{3}{4}$
3) $\frac{3}{2}$
4) $\frac{4}{3}$

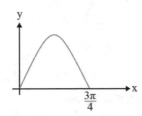

Figure 1.1 The graph of problem 1.9

1.10. Figure 1.2 illustrates a part of the function of $y = \cos\left(\left(ax + \frac{1}{2}\right)\pi\right)$. Determine the value of a.

Difficulty level   ● Easy   ○ Normal   ○ Hard
Calculation amount   ● Small   ○ Normal   ○ Large

1) $\frac{1}{2}$
2) $\frac{3}{2}$
3) $\frac{2}{3}$
4) $\frac{7}{4}$

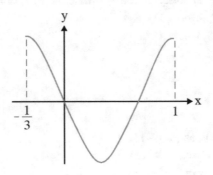

Figure 1.2 The graph of problem 1.10

1.11. Which one of the following choices is correct if $\alpha + \beta = 19\pi$.

Difficulty level   ● Easy   ○ Normal   ○ Hard
Calculation amount   ○ Small   ● Normal   ○ Large

1) $\sin(\alpha) = \sin(\beta)$
2) $\cos(\alpha) = \cos(\beta)$
3) $\tan(\alpha) = \tan(\beta)$
4) $\cot(\alpha) = \cot(\beta)$

1.12. Calculate the final value of the following equation.

$$\sin(5\pi + x) + \sin\left(x - \frac{\pi}{3}\right) + \sin\left(x + \frac{7\pi}{3}\right)$$

Difficulty level   ● Easy   ○ Normal   ○ Hard
Calculation amount   ○ Small   ● Normal   ○ Large

1) 0

2) $\sin\left(\frac{\pi}{3}\right)$

3) $2\sin\left(\frac{\pi}{3}\right)$

4) $-\sin\left(\frac{\pi}{3}\right)$

1.13. Calculate the value of $\cos(20°)$ if $\sin(50°) + \sin(10°) = m$.

Difficulty level   ○ Easy   ● Normal   ○ Hard

Calculation amount   ● Small   ○ Normal   ○ Large

1) $\frac{m}{2}$

2) m

3) $2m$

4) $\frac{2m}{3}$

1.14. Simplify and calculate the final value of the following term.

$$\frac{\left(1 + \tan^2\left(5°\right)\right)\sin\left(10°\right)}{\left(1 - \tan^2\left(5°\right)\right)\tan\left(10°\right)}$$

Difficulty level   ○ Easy   ● Normal   ○ Hard

Calculation amount   ● Small   ○ Normal   ○ Large

1) $\tan(15°)$

2) $\tan(25°)$

3) $\tan(35°)$

4) $\tan(45°)$

1.15. Which one of the following relations is correct if $\cot(\alpha) = m$ and $\cos(\alpha) = n$.

Difficulty level   ○ Easy   ● Normal   ○ Hard

Calculation amount   ● Small   ○ Normal   ○ Large

1) $m^2(1 + n^2) = n^2$

2) $m^2(1 - n^2) = n^2$

3) $m^2(2 + n^2) = 1$

4) $m^2(2 - n^2) = 1$

1.16. Determine the main period of $\sin(3x)\cos(5x) + 11$.

Difficulty level   ○ Easy   ● Normal   ○ Hard

Calculation amount   ● Small   ○ Normal   ○ Large

1) π

2) 2π

3) $\frac{2\pi}{3}$

4) $\frac{2\pi}{5}$

1.17. Calculate the value of $\text{arc}\left(\cos\left(\sin\left(\frac{4\pi}{3}\right)\right)\right)$.

Difficulty level   ○ Easy   ● Normal   ○ Hard

Calculation amount   ● Small   ○ Normal   ○ Large

1) $\frac{\pi}{6}$

2) $\frac{5\pi}{6}$

3) $\frac{\pi}{3}$

4) $-\frac{\pi}{6}$

1.18. Calculate the value of $\text{arc}\left(\sin\left(\sin\left(\frac{17\pi}{5}\right)\right)\right)$.

Difficulty level   ○ Easy   ● Normal   ○ Hard
Calculation amount   ● Small   ○ Normal   ○ Large

1) $\frac{2\pi}{5}$

2) $\frac{3\pi}{5}$

3) $-\frac{2\pi}{5}$

4) $-\frac{3\pi}{5}$

1.19. Calculate the value of $\text{arc}\left(\cos\left(\cos\left(\frac{19\pi}{5}\right)\right)\right)$.

Difficulty level   ○ Easy   ● Normal   ○ Hard
Calculation amount   ● Small   ○ Normal   ○ Large

1) $\frac{\pi}{5}$

2) $\frac{4\pi}{5}$

3) $-\frac{\pi}{5}$

4) $-\frac{4\pi}{5}$

1.20. Calculate the value of $\tan\left(2\text{arc}\left(\tan\left(\frac{1}{2}\right)\right)\right)$.

Difficulty level   ○ Easy   ● Normal   ○ Hard
Calculation amount   ● Small   ○ Normal   ○ Large

1) 1

2) $\frac{3}{4}$

3) $\frac{4}{3}$

4) $\frac{3}{5}$

1.21. Calculate the final value of $\sin\left(\text{arc}\left(\sin\left(\frac{3}{5}\right)\right)+\text{arc}\left(\tan\left(\frac{3}{4}\right)\right)\right)$.

Difficulty level   ○ Easy   ● Normal   ○ Hard
Calculation amount   ● Small   ○ Normal   ○ Large

1) $\frac{10}{13}$

2) $\frac{9}{13}$

3) $\frac{12}{35}$

4) $\frac{24}{25}$

1.22. Calculate the final value of $\text{arc}\left(\cot\left(-\frac{4}{3}\right)\right)-\text{arc}\left(\cot\left(\frac{3}{4}\right)\right)$.

Difficulty level   ○ Easy   ● Normal   ○ Hard
Calculation amount   ● Small   ○ Normal   ○ Large

1) π

2) $\frac{2\pi}{3}$

3) $\frac{\pi}{2}$

4) $\frac{4}{3}$

1.23. Calculate the final value of $\text{arc}\left(\tan\left(5\right)\right)+\text{arc}\left(\tan\left(\frac{3}{2}\right)\right)$.

Difficulty level   ○ Easy   ● Normal   ○ Hard
Calculation amount   ● Small   ○ Normal   ○ Large

1) $\frac{\pi}{4}$

2) $-\frac{\pi}{4}$

3) $\frac{3\pi}{4}$

4) $\frac{5\pi}{4}$

1.24. Calculate the final value of $\sin\left(\arc\left(\cos\left(\frac{3}{5}\right)\right)\right) + \cos\left(\arc\left(\sin\left(-\frac{4}{5}\right)\right)\right)$.

Difficulty level   ○ Easy   ● Normal   ○ Hard
Calculation amount   ● Small   ○ Normal   ○ Large

1) $\frac{7}{5}$
2) $-\frac{1}{5}$
3) $\frac{1}{5}$
4) $-\frac{7}{5}$

1.25. Figure 1.3 shows a unit circle. Which one of the choices shows the value of $\tan(\theta)$ and $\cot(\theta)$, respectively?

Difficulty level   ○ Easy   ● Normal   ○ Hard
Calculation amount   ● Small   ○ Normal   ○ Large

1) OA, OB
2) HA, HB
3) OA, AB
4) OB, BH

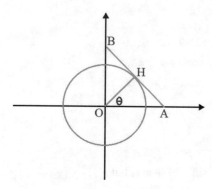

Figure 1.3 The graph of problem 1.25

1.26. Figure 1.4 illustrates a unit circle. Which one of the choices shows the value of $\sec(\theta)$?

Difficulty level   ○ Easy   ● Normal   ○ Hard
Calculation amount   ● Small   ○ Normal   ○ Large

1) HA
2) MB
3) OB
4) OM

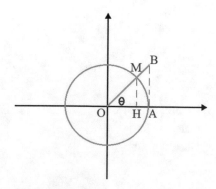

Figure 1.4 The graph of problem 1.26

1.27. Figure 1.5 illustrates a unit circle. Which one of the choices shows the value of csc(θ)?

1) *MB*
2) *OB*
3) *HC*
4) *OM*

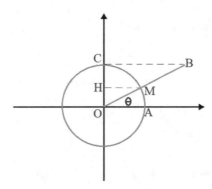

Figure 1.5 The graph of problem 1.27

1.28. Calculate the value of $\arc\left(\tan\left(\frac{2}{3}\right)\right) + \arc\left(\tan\left(\frac{1}{5}\right)\right)$.

1) $\frac{\pi}{6}$
2) $\frac{\pi}{4}$
3) $\frac{\pi}{3}$
4) $\frac{\pi}{2}$

1.29. Calculate the final value of the term below.

$$\arc(\tan(m)) + \arc\left(\tan\left(\frac{1}{m}\right)\right) + \arc(\cot(m)) + \arc(\cot(-m))$$

1) π or 2π
2) $\frac{\pi}{2}$ or $\frac{3\pi}{2}$
3) $\frac{3\pi}{2}$
4) $\frac{\pi}{2}$

1.30. Determine the range of x in the inequality below. Herein, x is an acute angle.

$$-1 \leq \cos(4x)\cos(2x) + \sin(4x)\sin(2x) \leq 0$$

1) $\left[\frac{\pi}{6}, \frac{3\pi}{8}\right]$
2) $\left[\frac{\pi}{8}, \frac{\pi}{4}\right]$
3) $\left[\frac{\pi}{6}, \frac{\pi}{3}\right]$
4) $\left[\frac{\pi}{4}, \frac{\pi}{2}\right]$

1.31. Calculate the value of tan(2y) if tan(x + y) = 5 and tan(x − y) = 7.

Difficulty level   ○ Easy   ● Normal   ○ Hard
Calculation amount   ● Small   ○ Normal   ○ Large

1) $\frac{1}{18}$

2) $-\frac{1}{18}$

3) $\frac{1}{36}$

4) $-\frac{1}{36}$

1.32. Simplify and calculate the value of the following term.

$$\frac{\sin\left(\frac{5\pi}{12}\right) + \cos\left(\frac{5\pi}{12}\right)}{\sin\left(\frac{5\pi}{12}\right) - \cos\left(\frac{5\pi}{12}\right)}$$

Difficulty level   ○ Easy   ● Normal   ○ Hard
Calculation amount   ● Small   ○ Normal   ○ Large

1) $\sqrt{3}$

2) $\frac{\sqrt{3}}{3}$

3) $-2\sqrt{3}$

4) $-\frac{\sqrt{3}}{3}$

1.33. Figure 1.6 illustrates a part of the function of $y = a \sin(b\pi x)$. Determine the value of $a + b$.

Difficulty level   ○ Easy   ● Normal   ○ Hard
Calculation amount   ● Small   ○ Normal   ○ Large

1) $\frac{4}{3}$

2) $\frac{5}{3}$

3) $\frac{7}{3}$

4) $\frac{8}{3}$

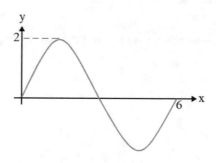

Figure 1.6 The graph of problem 1.33

1.34. Figure 1.7 illustrates a part of the function of $y = a \sin(b\pi x)$. Determine the value of $a \times b$.

Difficulty level   ○ Easy   ● Normal   ○ Hard
Calculation amount   ● Small   ○ Normal   ○ Large

1) −6

2) −3

3) $\frac{9}{2}$

4) 6

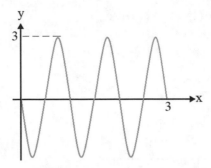

Figure 1.7 The graph of problem 1.34

1.35. Figure 1.8 illustrates a part of the function of $y = a \sin\left(\left(bx + \frac{1}{2}\right)\pi\right)$. Determine the value of $a \times b$.

Difficulty level   ○ Easy   ● Normal   ○ Hard
Calculation amount   ● Small   ○ Normal   ○ Large

1) 2
2) $\frac{5}{2}$
3) 3
4) $\frac{7}{2}$

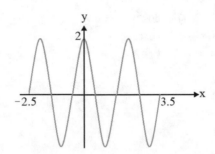

Figure 1.8 The graph of problem 1.35

1.36. Simplify and calculate the value of the following term.

$$\frac{\cos\left(5^\circ\right)\cos\left(10^\circ\right)\cos\left(20^\circ\right)}{\cos\left(50^\circ\right)}$$

Difficulty level   ○ Easy   ● Normal   ○ Hard
Calculation amount   ○ Small   ● Normal   ○ Large

1) $\frac{1}{4\cos\left(85^\circ\right)}$
2) $\frac{1}{8\cos\left(85^\circ\right)}$
3) $\frac{1}{8\sin\left(85^\circ\right)}$
4) $\frac{1}{4\sin\left(85^\circ\right)}$

1.37. Calculate the value of $\tan \left(\frac{x}{2}\right)$ if $\sin (x) + \cos (x) = \frac{7}{5}$.

Difficulty level   ○ Easy   ● Normal   ○ Hard
Calculation amount   ○ Small   ● Normal   ○ Large

1) 2 or 3
2) $\frac{1}{2}$ or $\frac{1}{3}$
3) 2 or $\frac{3}{5}$
4) 3 or $\frac{2}{5}$

1.38. Simplify and calculate the value of the following term.

$$\frac{\sin^4(\alpha) - \cos^4(\alpha)}{\sin(\alpha)\cos(\alpha)}$$

Difficulty level   ○ Easy   ● Normal   ○ Hard
Calculation amount   ○ Small   ● Normal   ○ Large

1) $2 \cot(2\alpha)$
2) $-2 \cot(2\alpha)$
3) $2 \tan(3\alpha)$
4) $-2 \tan(3\alpha)$

1.39. Calculate the value of $\cot^2(2\alpha)$ if $\sin^4(\alpha) + \cos^4(\alpha) = \frac{1}{2}$.

Difficulty level   ○ Easy   ● Normal   ○ Hard
Calculation amount   ○ Small   ● Normal   ○ Large

1) 1
2) 2
3) 0
4) 3

1.40. Calculate the value of the following relation for $x = \frac{3\pi}{8}$.

$$\sin^3(x)\cos(x) - \cos^3(x)\sin(x) + 3\sin^2(x)\cos^2(x)$$

Difficulty level   ○ Easy   ● Normal   ○ Hard
Calculation amount   ○ Small   ● Normal   ○ Large

1) $\frac{3}{8}$
2) $\frac{5}{8}$
3) $-\frac{5}{8}$
4) $-\frac{3}{8}$

1.41. Calculate the value of the following relation for $\alpha = \frac{\pi}{15}$.

$$\frac{\sin(2\alpha) + \sin(5\alpha) + \sin(8\alpha)}{\cos(2\alpha) + \cos(5\alpha) + \cos(8\alpha)}$$

Difficulty level   ○ Easy   ● Normal   ○ Hard
Calculation amount   ○ Small   ● Normal   ○ Large

1) $-\frac{\sqrt{3}}{3}$
2) $-\sqrt{3}$
3) $\sqrt{3}$
4) $\frac{\sqrt{3}}{3}$

1.42. Calculate the value of the following relation for $x = \frac{\pi}{12}$.

$$(\sin (x) - \cos (x) + 2)(\sin (x) - \cos (x) - 2)$$

Difficulty level   ○ Easy   ● Normal   ○ Hard
Calculation amount   ○ Small   ● Normal   ○ Large

1) $\frac{7}{2}$
2) $\frac{5}{2}$
3) $-\frac{5}{2}$
4) $-\frac{7}{2}$

1.43. Calculate the value of $4\sin^2(\alpha)\cos^2(\alpha)(\tan(\alpha) + \cot(\alpha))^2$.

Difficulty level   ○ Easy   ● Normal   ○ Hard
Calculation amount   ○ Small   ● Normal   ○ Large

1) 1
2) 2
3) 3
4) 4

1.44. Determine the number of roots of the equation below.

$$\sin(\pi x) \cos^2(\pi x) + \sin^2(\pi x) \cos(\pi x) = 0$$

Difficulty level   ○ Easy   ● Normal   ○ Hard
Calculation amount   ○ Small   ● Normal   ○ Large

1) 11
2) 12
3) 13
4) 14

1.45. Figure 1.9 illustrates a part of the function of $y = \frac{1}{2} + 2\cos(mx)$. Determine the value of the function for $x = \frac{16\pi}{3}$.

Difficulty level   ○ Easy   ● Normal   ○ Hard
Calculation amount   ○ Small   ● Normal   ○ Large

1) $-\frac{1}{2}$
2) $\frac{1}{2}$
3) 1
4) 0

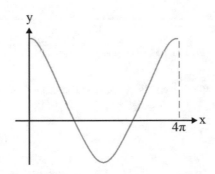

Figure 1.9 The graph of problem 1.45

1.46. Figure 1.10 shows a part of the function of $y = 1 + \sin(mx)$. Determine the value of the function for $x = \frac{7\pi}{6}$.

Difficulty level   ○ Easy   ● Normal   ○ Hard
Calculation amount   ○ Small   ● Normal   ○ Large

1) 0
2) $\frac{1}{2}$
3) 1
4) 2

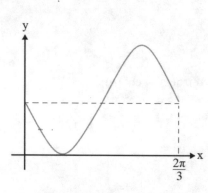

Figure 1.10 The graph of problem 1.46

1.47. Figure 1.11 shows a part of the function of $y = a - \sin(b\pi x)$. Determine the value of the function for $x = \frac{25}{3}$.

Difficulty level   ○ Easy   ● Normal   ○ Hard
Calculation amount   ○ Small   ● Normal   ○ Large

1) 2
2) $\frac{5}{2}$
3) 3
4) $\frac{7}{2}$

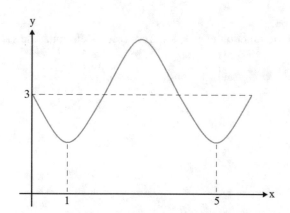

Figure 1.11 The graph of problem 1.47

1.48. Figure 1.12 shows the function of $y = a + b \cos\left(\frac{\pi}{2}x\right)$ for $0 < x < 4$. Determine the value of b.

Difficulty level   ○ Easy   ● Normal   ○ Hard
Calculation amount   ○ Small   ● Normal   ○ Large

1) -2
2) -1
3) 1
4) 2

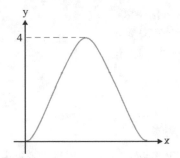

Figure 1.12 The graph of problem 1.48

1.49. Figure 1.13 shows the function of $y = 1 + a \sin(b\pi x)$ for $0 < x < \frac{4}{3}$. Determine the value of $a + b$.

Difficulty level   ○ Easy   ● Normal   ○ Hard
Calculation amount   ○ Small   ● Normal   ○ Large

1) 3
2) 4
3) 5
4) 6

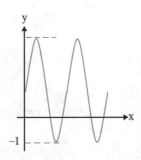

Figure 1.13 The graph of problem 1.49

1.50. Figure 1.14 shows a part the function of $y = a - 2\cos\left(bx + \frac{\pi}{2}\right)$. Determine the value of $a + b$.

Difficulty level   ○ Easy   ● Normal   ○ Hard
Calculation amount   ○ Small   ● Normal   ○ Large

1) $\frac{1}{2}$
2) 1
3) $\frac{3}{2}$
4) 2

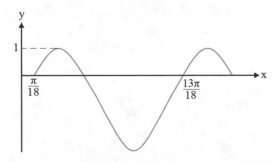

Figure 1.14 The graph of problem 1.50

1.51. Calculate the value of $\cos(25° - \alpha)$ if $\tan\left(\alpha + 20°\right) = \frac{3}{4}$.

Difficulty level　　　○ Easy　● Normal　○ Hard
Calculation amount　○ Small　● Normal　○ Large

1) 5
2) 6
3) 7
4) 8

1.52. Calculate the value of $\tan\left(\frac{\pi}{4} + \alpha\right)$ assuming that α is an acute angle and $\sin(\alpha) = \frac{3}{5}$.

Difficulty level　　　○ Easy　● Normal　○ Hard
Calculation amount　○ Small　● Normal　○ Large

1) -7
2) $-\frac{1}{7}$
3) $\frac{1}{7}$
4) 7

1.53. Calculate the value of $\tan\left(\frac{\pi}{4} - \alpha\right)$ if $\tan\left(\frac{\pi}{2} - \alpha\right) = \frac{2}{3}$.

Difficulty level　　　○ Easy　● Normal　○ Hard
Calculation amount　○ Small　● Normal　○ Large

1) $-\frac{1}{3}$
2) $-\frac{1}{5}$
3) $\frac{1}{5}$
4) $\frac{1}{3}$

1.54. Calculate the value of $\tan(2a)$ while we know that $\tan(a+b) = \frac{2}{5}$ and $\tan(a-b) = \frac{3}{7}$.

Difficulty level　　　○ Easy　● Normal　○ Hard
Calculation amount　○ Small　● Normal　○ Large

1) $-\frac{1}{3}$
2) $-\frac{1}{2}$
3) 3
4) 1

1.55. Calculate the value of $\tan(x)$ if we have:

$$\frac{\sin\left(x - \frac{\pi}{4}\right)}{\cos\left(x - \frac{\pi}{4}\right)} = 2$$

Difficulty level　　　○ Easy　● Normal　○ Hard
Calculation amount　○ Small　● Normal　○ Large

1) -3
2) $\frac{1}{3}$
3) $\frac{2}{3}$
4) 3

1.56. Calculate the value of $(1 + \tan(\alpha))(1 + \tan(\beta))$ if $\alpha + \beta = \frac{\pi}{4}$.

   Difficulty level   ○ Easy   ● Normal   ○ Hard
   Calculation amount   ○ Small   ● Normal   ○ Large

   1) -2
   2) 2
   3) $\frac{1}{3}$
   4) $-\frac{1}{2}$

1.57. Calculate the value of $\tan\left(\frac{\pi}{4} + \alpha\right) - \tan\left(\frac{\pi}{4} - \alpha\right)$.

   Difficulty level   ○ Easy   ● Normal   ○ Hard
   Calculation amount   ○ Small   ● Normal   ○ Large

   1) $2\tan(2\alpha)$
   2) $2\cos(2\alpha)$
   3) 0
   4) $2\sin(2\alpha)$

1.58. Calculate the value of $\tan(2\alpha)$ if $\tan\left(\frac{\pi}{4} - \alpha\right) = \frac{1}{5}$.

   Difficulty level   ○ Easy   ● Normal   ○ Hard
   Calculation amount   ○ Small   ● Normal   ○ Large

   1) 1.5
   2) 1.8
   3) 2.4
   4) 2.5

1.59. Calculate the value of $\tan(2\alpha - \beta)$ if $\tan(\alpha) = 2$ and $\tan(\beta) = \frac{1}{3}$.

   Difficulty level   ○ Easy   ● Normal   ○ Hard
   Calculation amount   ○ Small   ● Normal   ○ Large

   1) -3
   2) -2
   3) 0.5
   4) 3

1.60. Determine the common solution of the equation of $\cos(3x) + \cos(x) = 0$ assuming $\cos(x) \neq 0$.

   Difficulty level   ○ Easy   ● Normal   ○ Hard
   Calculation amount   ○ Small   ● Normal   ○ Large

   1) $\frac{k\pi}{2} + \frac{\pi}{4}$
   2) $\frac{k\pi}{2} + \frac{\pi}{8}$
   3) $k\pi - \frac{\pi}{4}$
   4) $k\pi + \frac{\pi}{4}$

1.61. Calculate the sum of the positive acute roots of the equation of $\tan(4x) = \cot(x)$.

   Difficulty level   ○ Easy   ● Normal   ○ Hard
   Calculation amount   ○ Small   ● Normal   ○ Large

   1) $\frac{2\pi}{5}$
   2) $\frac{4\pi}{5}$
   3) $\frac{3\pi}{5}$
   4) $\frac{\pi}{5}$

1.62. Determine the common solution of the equation of $2\sin^2(x) + 3\cos(x) = 0$

    Difficulty level   ○ Easy   ● Normal   ○ Hard

    Calculation amount   ○ Small   ● Normal   ○ Large

    1) $2k\pi \pm \frac{2\pi}{3}$

    2) $2k\pi \pm \frac{\pi}{3}$

    3) $2k\pi \pm \frac{5\pi}{6}$

    4) $k\pi - \frac{\pi}{3}$

1.63. Determine the common solution of the equation of $2\sin^2(x) = 3\cos(x)$.

    Difficulty level   ○ Easy   ● Normal   ○ Hard

    Calculation amount   ○ Small   ● Normal   ○ Large

    1) $k\pi \pm \frac{\pi}{6}$

    2) $k\pi \pm \frac{\pi}{3}$

    3) $2k\pi \pm \frac{\pi}{6}$

    4) $2k\pi \pm \frac{\pi}{3}$

1.64. Two lines with the equations of $x\tan(\alpha) + y\cot(\alpha) = 2$ and $x\tan(\alpha) - y\cot(\alpha) = 1$ are intersecting each other at point M. By changing the value of α, what is the position equation of the point?

    Difficulty level   ○ Easy   ○ Normal   ● Hard

    Calculation amount   ● Small   ○ Normal   ○ Large

    1) $y = \frac{1}{x}$

    2) $y = \frac{3}{x}$

    3) $y = \frac{1}{4x}$

    4) $y = \frac{3}{4x}$

1.65. What is the position equation of the point of $(2 - 3\sin(\alpha), 1 + 4\cos(\alpha))$ if the value of α changes?

    Difficulty level   ○ Easy   ○ Normal   ● Hard

    Calculation amount   ● Small   ○ Normal   ○ Large

    1) Circle

    2) Ellipse

    3) Parabola

    4) Hyperbola

1.66. What is the position equation of the point of $(2 - 5\cos(\alpha), 4)$ if the value of α changes?

    Difficulty level   ○ Easy   ○ Normal   ● Hard

    Calculation amount   ● Small   ○ Normal   ○ Large

    1) A horizontal line

    2) A vertical line

    3) A horizontal line segment

    4) A vertical line segment

1.67. Calculate the value of y if $2\cos(x - y) + 3\sin(x + y) = 5$ and $0 < x, y < 2\pi$.

    Difficulty level   ○ Easy   ○ Normal   ● Hard

    Calculation amount   ● Small   ○ Normal   ○ Large

    1) $\frac{\pi}{3}$ or $\frac{2\pi}{3}$

    2) $\frac{\pi}{4}$ or $\frac{5\pi}{4}$

    3) $\frac{\pi}{6}$ or $\frac{5\pi}{6}$

    4) $\frac{\pi}{2}$ or $\frac{3\pi}{2}$

1.68. Calculate the value of m if $\tan(\alpha) \neq \tan(\beta)$, $\alpha + \beta = \frac{\pi}{4}$ and α and β are the two roots of the equation below.

$$\tan^2(x) + (m+2)\tan(x) + 2m - 2 = 0$$

Difficulty level   ○ Easy   ○ Normal   ● Hard
Calculation amount   ● Small   ○ Normal   ○ Large
1) 1
2) 3
3) 5
4) 7

1.69. Calculate the final value of the following relation.

$$\frac{\sin^6(\alpha) + \cos^6(\alpha) + 3\sin^2(\alpha)\cos^2(\alpha)}{\sin^4(\alpha) + \cos^4(\alpha) + 2\sin^2(\alpha)\cos^2(\alpha)}$$

Difficulty level   ○ Easy   ○ Normal   ● Hard
Calculation amount   ● Small   ○ Normal   ○ Large
1) $\sin^2(\alpha)$
2) $\cos^2(\alpha)$
3) $\sin^2(\alpha) - \cos^2(\alpha)$
4) 1

1.70. Calculate the final value of the relation below.

$$\frac{\sin\left(135^\circ\right)\cos\left(210^\circ\right) + \cos\left(135^\circ\right)\sin\left(420^\circ\right)}{\tan\left(210^\circ\right)\cot\left(420^\circ\right) + \cot\left(120^\circ\right)\tan\left(330^\circ\right)}$$

Difficulty level   ○ Easy   ● Normal   ○ Hard
Calculation amount   ○ Small   ● Normal   ○ Large
1) $-\frac{\sqrt{6}}{4}$
2) $-\frac{3\sqrt{6}}{4}$
3) $-\frac{\sqrt{6}}{2}$
4) $-\frac{3\sqrt{6}}{2}$

1.71. Calculate the final value of $(1 + \cot(x))(1 + \cot(y))$ if $x + y = k\pi + \frac{\pi}{4}$.
Difficulty level   ○ Easy   ○ Normal   ● Hard
Calculation amount   ○ Small   ● Normal   ○ Large
1) $\tan(x)\tan(y)$
2) $2\tan(x)\tan(y)$
3) $\cot(x)\cot(y)$
4) $2\cot(x)\cot(y)$

1.72. Determine the common solution of the equation below.

$$(\sin(x) - \tan(x))\tan\left(\frac{3\pi}{2} - x\right) = \cos\left(\frac{4\pi}{3}\right)$$

Difficulty level   ○ Easy   ○ Normal   ● Hard
Calculation amount   ○ Small   ● Normal   ○ Large

1) $k\pi - \frac{\pi}{6}$

2) $k\pi + \frac{\pi}{3}$

3) $2k\pi \pm \frac{\pi}{3}$

4) $2k\pi \pm \frac{\pi}{6}$

1.73. Calculate the sum of the roots of the equation below for $x \in [0, \pi]$.

$$\sin(2x)(\sin(x) + \cos(x)) = \cos(2x)(\cos(x) - \sin(x))$$

Difficulty level   ○ Easy   ○ Normal   ● Hard
Calculation amount   ○ Small   ● Normal   ○ Large

1) $\frac{3\pi}{4}$

2) $\frac{5\pi}{4}$

3) $\frac{3\pi}{2}$

4) $\frac{7\pi}{4}$

1.74. Determine the common solution of the equation of $\sqrt{2}\sin\left(\frac{\pi}{4} - x\right) = 1 + \sin\left(\frac{5\pi}{2} + x\right)$.

Difficulty level   ○ Easy   ○ Normal   ● Hard
Calculation amount   ○ Small   ● Normal   ○ Large

1) $k\pi + \frac{\pi}{2}$

2) $2k\pi - \frac{\pi}{4}$

3) $2k\pi - \frac{\pi}{2}$

4) $2k\pi + \frac{\pi}{2}$

1.75. Which one of the following choices shows one of the common solutions of the equation of $\cos(2x) + \sqrt{3}\sin(2x) = 1$.

Difficulty level   ○ Easy   ○ Normal   ● Hard
Calculation amount   ○ Small   ● Normal   ○ Large

1) $k\pi - \frac{\pi}{6}$

2) $k\pi - \frac{\pi}{3}$

3) $k\pi + \frac{\pi}{6}$

4) $k\pi + \frac{\pi}{3}$

Reference

1. Rahmani-Andebili, M. (2020). Precalculus: Practice problems, methods, and solutions, Springer Nature, 2020.

Solutions of Problems: Trigonometric Equations and Identities

2

Abstract

In this chapter, the problems of the first chapter are fully solved, in detail, step-by-step, and with different methods. The subjects include trigonometric equations, trigonometric identities, domain, range, period, sine and cosine identities, tangent and cotangent identities, half angle formulas, reciprocal identities, Pythagorean identities, sum and difference to product formulas, product to sum formulas, even and odd formulas, periodic formulas, sum to product formulas, double angle formulas, degrees to radians formulas, cofunction formulas, unit circle, inverse trigonometric functions, inverse properties, alternate notation, and domain and range of inverse trigonometric functions.

2.1. From trigonometry, we know that [1]

$$\tan(\theta) = \frac{1}{\cot(\theta)}$$

$$\tan(2\theta) = \frac{2\tan(\theta)}{1 - \tan^2(\theta)}$$

Based on the information given in the problem:

$$\cot(\theta) = 5 \Rightarrow \tan(\theta) = \frac{1}{5}$$

Therefore,

$$\tan(2\theta) = \frac{2\tan(\theta)}{1 - \tan^2(\theta)} = \frac{2 \times \frac{1}{5}}{1 - \left(\frac{1}{5}\right)^2} = \frac{\frac{2}{5}}{\frac{24}{25}} = \frac{5}{12}$$

Choice (1) is the answer.

2.2. From trigonometry, we know that:

$$\tan(\alpha + n\pi) = \tan(\alpha), \forall n \in \mathbb{Z}$$

$$\tan(-\alpha) = -\tan(\alpha)$$

Therefore,

$$\tan\left(-2100^{\circ}\right) = -\tan\left(2100^{\circ}\right) = -\tan\left(12 \times 180 - 60^{\circ}\right) = -\tan\left(-60^{\circ}\right) = \tan\left(60^{\circ}\right) = \sqrt{3}$$

Choice (1) is the answer.

2.3. From trigonometry, we know that:

$$1 + \cos\left(\theta\right) = 2\cos^{2}\left(\frac{\theta}{2}\right)$$

$$\sin\left(\theta\right) = 2\sin\left(\frac{\theta}{2}\right)\cos\left(\frac{\theta}{2}\right)$$

$$\cot\left(\theta\right) = \frac{\cos\left(\theta\right)}{\sin\left(\theta\right)}$$

Therefore,

$$\frac{1 + \cos\left(40^{\circ}\right)}{\sin\left(40^{\circ}\right)} = \frac{2\cos^{2}\left(20^{\circ}\right)}{2\sin\left(20^{\circ}\right)\cos\left(20^{\circ}\right)} = \frac{\cos\left(20^{\circ}\right)}{\sin\left(20^{\circ}\right)} = \cot\left(20^{\circ}\right)$$

Choice (4) is the answer.

2.4. For the given range of α, we can conclude that:

$$\frac{\pi}{6} \le \alpha \le \frac{2\pi}{3} \Rightarrow \frac{1}{2} \le \sin\left(\alpha\right) \le 1$$

Therefore, based on the given information, i.e., $\sin\left(\alpha\right) = \frac{3m-1}{4}$, we can write:

$$\frac{1}{2} \le \frac{3m-1}{4} \le 1 \Rightarrow 1 \le m \le \frac{5}{3}$$

Choice (4) is the answer.

2.5. For the given range of x, we can conclude that:

$$-\frac{\pi}{3} \le x \le \frac{\pi}{3} \Rightarrow \frac{1}{2} \le \cos\left(x\right) \le 1$$

Therefore, based on the given information, i.e., $\cos\left(\alpha\right) = \frac{2m-1}{6}$, we can write:

$$\frac{1}{2} \le \frac{2m-1}{6} \le 1 \Rightarrow 2 \le m \le \frac{7}{2}$$

Choice (1) is the answer.

2.6. From trigonometry, we know that:

$$f_{1}(x) = \cos^{2n}(ax), \forall n \in \mathbb{Z} \Rightarrow T_{1} = \frac{\pi}{|a|}$$

$$f_{2}(x) = \cos^{2n+1}(ax), \forall n \in \mathbb{Z} \Rightarrow T_{2} = \frac{2\pi}{|a|}$$

Therefore,

$$f_1(x) = \cos^2(x) \Rightarrow T_1 = \frac{\pi}{1} = \pi$$

$$f_2(x) = -5\cos\left(\frac{2x}{3}\right) \Rightarrow T_2 = \frac{2\pi}{\frac{2}{3}} = 3\pi$$

The main period of the given expression is the least common multiple (LCM) of the main periods of the terms, as can be seen in the following.

$$\Rightarrow T = \text{LCM}(\pi, 3\pi) = 3\pi$$

Choice (3) is the answer.

2.7. From trigonometry, we know that:

$$f_1(x) = \sin^{2n}(ax), \forall n \in \mathbb{Z} \Rightarrow T_1 = \frac{\pi}{|a|}$$

$$f_2(x) = \cos^{2n+1}(ax), \forall n \in \mathbb{Z} \Rightarrow T_2 = \frac{2\pi}{|a|}$$

Therefore,

$$f_1(x) = \sin^4\left(\frac{3x}{5}\right) \Rightarrow T_1 = \frac{\pi}{\frac{3}{5}} = \frac{5\pi}{3}$$

$$f_2(x) = \cos^3\left(\frac{2x}{3}\right) \Rightarrow T_2 = \frac{2\pi}{\frac{2}{3}} = 3\pi$$

The main period of the given expression is the least common multiple (LCM) of the main periods of the terms as follows.

$$\Rightarrow T = \text{LCM}\left(\frac{5\pi}{3}, 3\pi\right) = 15\pi$$

Choice (3) is the answer.

2.8. From trigonometry, we can know that:

$$f_1(x) = \sin^{2n}(ax), \forall n \in \mathbb{Z} \Rightarrow T_1 = \frac{\pi}{|a|}$$

$$f_2(x) = \cos^{2n+1}(ax), \forall n \in \mathbb{Z} \Rightarrow T_2 = \frac{2\pi}{|a|}$$

Therefore,

$$f_1(x) = \sin^4\left(\frac{\pi x}{3}\right) \Rightarrow T_1 = \frac{\pi}{\frac{\pi}{3}} = 3$$

$$f_2(x) = \cos(\pi x) \Rightarrow T_2 = \frac{2\pi}{\pi} = 2$$

The main period of the given expression is the least common multiple (LCM) of the main periods of the terms, as can be seen in the following.

$$\Rightarrow T = \text{LCM}(3, 2) = 6$$

2.9. From trigonometry, we know that:

$$y = \sin(kx) \Rightarrow T = \frac{2\pi}{|k|}$$

Therefore,

$$\Rightarrow \frac{3\pi}{4} = \frac{1}{2}\left(\frac{2\pi}{|k|}\right) \Rightarrow |k| = \frac{4}{3} \Rightarrow k = \pm\frac{4}{3}$$

Based on the graph and the function, the positive value of k is acceptable.

$$\Rightarrow k = \frac{4}{3}$$

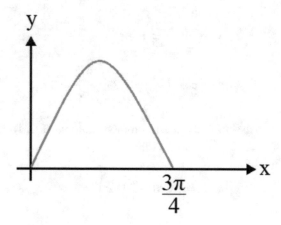

Figure 2.1 The graph of the solution of problem 2.9

2.10. From trigonometry, we know that:

$$y = \cos\left(\pi a x + \frac{\pi}{2}\right) = -\sin(\pi a x)$$

$$y = \sin(mx) \Rightarrow T = \frac{2\pi}{|m|}$$

Therefore,

$$\Rightarrow 1 - \left(-\frac{1}{3}\right) = \frac{2\pi}{|\pi a|} \Rightarrow \frac{4}{3} = \frac{2}{|a|} \Rightarrow a = \pm\frac{3}{2}$$

Based on the graph and $y = -\sin(\pi a x)$, the positive value of a is acceptable.

$$\Rightarrow a = \frac{3}{2}$$

Choice (2) is the answer.

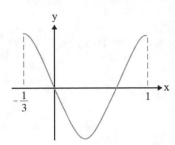

Figure 2.2 The graph of the solution of problem 2.10

2.11. From trigonometry, we know that:

$$\sin(\alpha + 2n\pi) = \sin(\alpha), \forall n \in \mathbb{Z}$$

$$\cos(\alpha + 2n\pi) = \cos(\alpha), \forall n \in \mathbb{Z}$$

$$\tan(\alpha + n\pi) = \tan(\alpha), \forall n \in \mathbb{Z}$$

$$\cot(\alpha + n\pi) = \cot(\alpha), \forall n \in \mathbb{Z}$$

$$\sin(\pi - \alpha) = \sin(\alpha)$$

$$\cos(\pi - \alpha) = -\cos(\alpha)$$

$$\tan(-\alpha) = -\tan(\alpha)$$

$$\cot(-\alpha) = -\cot(\alpha)$$

Based on the information given in the problem, we have:

$$\alpha + \beta = 19\pi \Rightarrow \alpha = 19\pi - \beta$$

Therefore,

$$\sin(\alpha) = \sin(19\pi - \beta) = \sin(\pi - \beta) = \sin(\beta)$$

$$\cos(\alpha) = \cos(19\pi - \beta) = \cos(\pi - \beta) = -\cos(\beta)$$

$$\tan(\alpha) = \tan(19\pi - \beta) = \tan(-\beta) = -\tan(\beta)$$

$$\cot(\alpha) = \cot(19\pi - \beta) = \cot(-\beta) = -\cot(\beta)$$

Choice (1) is the answer.

2.12. From trigonometry, we know that:

$$\sin(\alpha + 2n\pi) = \sin(\alpha), \forall n \in \mathbb{Z}$$

$$\cos(\alpha + 2n\pi) = \cos(\alpha), \forall n \in \mathbb{Z}$$

$$\sin(\alpha + \pi) = -\sin(\alpha)$$

$$\sin(\alpha) + \sin(\beta) = 2\sin\left(\frac{\alpha+\beta}{2}\right)\cos\left(\frac{\alpha-\beta}{2}\right)$$

Therefore,

$$\sin(5\pi + x) + \sin\left(x - \frac{\pi}{3}\right) + \sin\left(x + \frac{7\pi}{3}\right) = \sin(x+\pi) + \sin\left(x - \frac{\pi}{3}\right) + \sin\left(x + \frac{\pi}{3}\right)$$

$$= -\sin(x) + 2\sin(x)\cos\left(\frac{\pi}{3}\right) = -\sin(x) + \sin(x) = 0$$

Choice (1) is the answer.

2.13. From trigonometry, we know that:

$$\sin(\alpha) + \sin(\beta) = 2\sin\left(\frac{\alpha+\beta}{2}\right)\cos\left(\frac{\alpha-\beta}{2}\right)$$

Therefore,

$$\sin(50°) + \sin(10°) = m \Rightarrow 2\sin(30°)\cos(20°) = m \Rightarrow 2 \times \frac{1}{2}\cos(20°) = m \Rightarrow \cos(20°) = m$$

Choice (2) is the answer.

2.14. From trigonometry, we know that:

$$\cos(\theta) = \frac{1 - \tan^2\left(\frac{\theta}{2}\right)}{1 + \tan^2\left(\frac{\theta}{2}\right)}$$

$$\tan(\theta) = \frac{\sin(\theta)}{\cos(\theta)}$$

$$\tan(45°) = 1$$

Therefore,

$$\frac{(1 + \tan^2(5°))\sin(10°)}{(1 - \tan^2(5°))\tan(10°)} = \frac{1}{\cos(10°)} \times \frac{\sin(10°)}{\frac{sin(10°)}{\cos(10°)}} = 1 = \tan(45°)$$

Choice (4) is the answer.

2.15. From trigonometry, we know that:

$$1 + \tan^2(\alpha) = \frac{1}{\cos^2(\alpha)}$$

$$\tan(\alpha) = \frac{1}{\cot(\alpha)}$$

Based on the information given in the problem, we have:

$$\cot(\alpha) = m$$

$$\cos(\alpha) = n$$

Therefore,

$$\frac{1}{\cos^2(\alpha)} = 1 + \tan^2(\alpha) = 1 + \frac{1}{\cot^2(\alpha)} \Rightarrow \frac{1}{n^2} = 1 + \frac{1}{m^2} \xrightarrow{\times m^2 n^2} m^2 = m^2 n^2 + n^2$$

$$\Rightarrow m^2(1 - n^2) = n^2$$

Choice (2) is the answer.

2.16. From trigonometry, we can know that:

$$\sin^{2n+1}(ax), \forall n \in \mathbb{Z} \Rightarrow T = \frac{2\pi}{|a|}$$

$$\sin(\alpha)\cos(\beta) = \frac{1}{2}(\sin(\alpha+\beta) + \sin(\alpha-\beta))$$

We need to change the product expression to the summation one, as follows.

$$y = \sin(3x)\cos(5x) + 11 \Rightarrow y = \frac{1}{2}\sin(8x) - \frac{1}{2}\sin(2x) + 11$$

Then,

$$\frac{1}{2}\sin(8x) \Rightarrow T_1 = \frac{2\pi}{8} = \frac{\pi}{4}$$

$$-\frac{1}{2}\sin(2x) \Rightarrow T_2 = \frac{2\pi}{2} = \pi$$

The main period of the given expression is the least common multiple (LCM) of the main periods of the terms, as is presented in the following.

$$\Rightarrow T = \text{LCM}\left(\frac{\pi}{4}, \pi\right) = \pi$$

Choice (1) is the answer.

2.17. From trigonometry, we know that:

$$\sin\left(\frac{4\pi}{3}\right) = \sin\left(\pi + \frac{\pi}{3}\right) = -\sin\left(\frac{\pi}{3}\right) = -\frac{\sqrt{3}}{2}$$

$$\mathrm{arc}(\cos(-\alpha)) = \pi - \mathrm{arc}(\cos(\alpha))$$

$$\mathrm{arc}\left(\cos\left(\frac{\sqrt{3}}{2}\right)\right) = \frac{\pi}{6}$$

Therefore,

$$\mathrm{arc}\left(\cos\left(\sin\left(\frac{4\pi}{3}\right)\right)\right) = \mathrm{arc}\left(\cos\left(-\frac{\sqrt{3}}{2}\right)\right) = \pi - \mathrm{arc}\left(\cos\left(\frac{\sqrt{3}}{2}\right)\right) = \pi - \frac{\pi}{6} = \frac{5\pi}{6}$$

Choice (2) is the answer.

2.18. From trigonometry, we know that:

$$\sin\left(\frac{17\pi}{5}\right) = \sin\left(4\pi - \frac{3\pi}{5}\right) = \sin\left(-\frac{3\pi}{5}\right)$$

$$\sin\left(-\frac{3\pi}{5}\right) \triangleq \alpha \Rightarrow \mathrm{arc}(\sin(\alpha)) = -\frac{3\pi}{5}$$

Therefore,

$$\mathrm{arc}\left(\sin\left(\sin\left(\frac{17\pi}{5}\right)\right)\right) = \mathrm{arc}\left(\sin\left(\sin\left(-\frac{3\pi}{5}\right)\right)\right) = \mathrm{arc}(\sin(\alpha)) = -\frac{3\pi}{5}$$

Choice (4) is the answer.

2.19. From trigonometry, we know that:

$$\cos\left(\frac{19\pi}{5}\right) = \cos\left(4\pi - \frac{\pi}{5}\right) = \cos\left(-\frac{\pi}{5}\right)$$

$$\cos\left(\frac{\pi}{5}\right) \triangleq \alpha \Rightarrow \mathrm{arc}(\cos(\alpha)) = \frac{\pi}{5}$$

Therefore,

$$\mathrm{arc}\left(\cos\left(\cos\left(\frac{19\pi}{5}\right)\right)\right) = \mathrm{arc}\left(\cos\left(\cos\left(-\frac{\pi}{5}\right)\right)\right) = \mathrm{arc}\left(\cos\left(\cos\left(\frac{\pi}{5}\right)\right)\right) = \mathrm{arc}(\cos(\alpha)) = \frac{\pi}{5}$$

Choice (1) is the answer.

2.20. From trigonometry, we know that:

$$\tan(2\alpha) = \frac{2\tan(\alpha)}{1 - \tan^2(\alpha)}$$

$$\mathrm{arc}\left(\tan\left(\frac{1}{2}\right)\right) \triangleq \alpha \Rightarrow \tan(\alpha) = \frac{1}{2}$$

Therefore,

$$\tan\left(2\text{arc}\left(\tan\left(\frac{1}{2}\right)\right)\right) = \tan(2\alpha) = \frac{2\tan(\alpha)}{1-\tan^2(\alpha)} = \frac{2\times\frac{1}{2}}{1-\left(\frac{1}{2}\right)^2} = \frac{1}{\frac{3}{4}} = \frac{4}{3}$$

Choice (3) is the answer.

2.21. From trigonometry, we know that:

$$\sin^2(\alpha) + \cos^2(\alpha) = 1$$

$$1 + \tan^2(\alpha) = \frac{1}{\cos^2(\alpha)}$$

Therefore,

$$\text{arc}\left(\sin\left(\frac{3}{5}\right)\right) \triangleq \alpha \Rightarrow \sin(\alpha) = \frac{3}{5} \Rightarrow \cos(\alpha) = \frac{4}{5}$$

$$\text{arc}\left(\tan\left(\frac{3}{4}\right)\right) \triangleq \beta \Rightarrow \tan(\beta) = \frac{3}{4} \Rightarrow \cos(\beta) = \frac{4}{5} \Rightarrow \sin(\beta) = \frac{3}{5}$$

$$\Rightarrow \sin\left(\text{arc}\left(\sin\left(\frac{3}{5}\right)\right) + \text{arc}\left(\tan\left(\frac{3}{4}\right)\right)\right) = \sin(\alpha+\beta) = \sin(\alpha)\cos(\beta) + \cos(\alpha)\sin(\beta)$$

$$= \frac{3}{5} \times \frac{4}{5} + \frac{4}{5} \times \frac{3}{5} = \frac{24}{25}$$

Choice (4) is the answer.

2.22. From trigonometry, we know that:

$$\text{arc}(\cot(-\alpha)) = \pi - \text{arc}(\cot(\alpha))$$

$$\text{arc}(\cot(\alpha)) + \text{arc}\left(\cot\left(\frac{1}{\alpha}\right)\right) = \frac{\pi}{2}$$

Therefore,

$$\text{arc}\left(\cot\left(-\frac{4}{3}\right)\right) - \text{arc}\left(\cot\left(\frac{3}{4}\right)\right) = \pi - \text{arc}\left(\cot\left(\frac{4}{3}\right)\right) - \text{arc}\left(\cot\left(\frac{3}{4}\right)\right)$$

$$= \pi - \left(\text{arc}\left(\cot\left(\frac{4}{3}\right)\right) + \text{arc}\left(\cot\left(\frac{3}{4}\right)\right)\right) = \pi - \frac{\pi}{2} = \frac{\pi}{2}$$

Choice (3) is the answer.

2.23. From trigonometry, we know that:

$$\tan(\alpha+\beta) = \frac{\tan(\alpha) + \tan(\beta)}{1 - \tan(\alpha)\tan(\beta)}$$

$$\text{arc}(\tan(5)) \triangleq \alpha \Rightarrow \tan(\alpha) = 5$$

$$\text{arc}\left(\tan\left(\frac{3}{2}\right)\right) \triangleq \beta \Longrightarrow \tan\left(\beta\right) = \frac{3}{2}$$

Therefore,

$$\tan\left(\text{arc}\left(\tan\left(5\right)\right) + \text{arc}\left(\tan\left(\frac{3}{2}\right)\right)\right) = \tan\left(\alpha + \beta\right) = \frac{\tan\left(\alpha\right) + \tan\left(\beta\right)}{1 - \tan\left(\alpha\right)\tan\left(\beta\right)} = \frac{5 + \frac{3}{2}}{1 - \frac{15}{2}} = \frac{\frac{13}{2}}{-\frac{13}{2}} = -1$$

$$\text{arc}\left(\tan\left(5\right)\right) + \text{arc}\left(\tan\left(\frac{3}{2}\right)\right) = \tan^{-1}(-1) = \frac{3\pi}{4}$$

Choice (3) is the answer.

2.24. From trigonometry, we know that:

$$\sin^2(\alpha) + \cos^2(\alpha) = 1$$

$$\text{arc}\left(\cos\left(\frac{3}{5}\right)\right) \triangleq \alpha \Longrightarrow \cos\left(\alpha\right) = \frac{3}{5}$$

$$\text{arc}\left(\sin\left(-\frac{4}{5}\right)\right) \triangleq \beta \Longrightarrow \sin\left(\beta\right) = -\frac{4}{5}$$

Therefore,

$$\sin\left(\text{arc}\left(\cos\left(\frac{3}{5}\right)\right)\right) + \cos\left(\text{arc}\left(\sin\left(-\frac{4}{5}\right)\right)\right) = \sin\left(\alpha\right) + \cos\left(\beta\right) = \sqrt{1 - \left(\frac{3}{5}\right)^2} + \sqrt{1 - \left(-\frac{4}{5}\right)^2}$$

$$= \frac{4}{5} + \frac{3}{5} = \frac{7}{5}$$

Choice (1) is the answer.

2.25. From trigonometry, we know that:

$$\cot\left(\theta\right) = \tan\left(\frac{\pi}{2} - \theta\right)$$

Based on the definition of $\tan(\theta)$ and $\cot(\theta)$, we can write:

$$\tan\left(\theta\right) = \frac{Opposite\ for\ \theta}{Adjacent\ for\ \theta} = \frac{HA}{OH} = \frac{HA}{1} = HA$$

$$\cot\left(\theta\right) = \tan\left(\widehat{HOB}\right) = \frac{Opposite\ for\ \widehat{HOB}}{Adjacent\ for\ \widehat{HOB}} = \frac{HB}{OH} = \frac{HB}{1} = HB$$

Choice (2) is the answer.

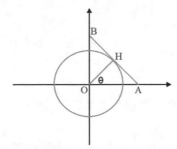

Figure 2.3 The graph of the solution of problem 2.25

2.26. Based on the definition of $\sec(\theta)$ and $\cos(\theta)$, we can write:

$$\sec(\theta) = \frac{1}{\cos(\theta)} = \frac{1}{\frac{Adjacent\ for\ \theta}{Hypotenuse\ for\ \theta}} = \frac{1}{\frac{OA}{OB}} = \frac{1}{\frac{1}{OB}} = OB$$

Choice (3) is the answer.

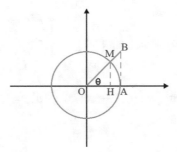

Figure 2.4 The graph of the solution of problem 2.26

2.27. From trigonometry, we know that:

$$\sin(\theta) = \cos\left(\frac{\pi}{2} - \theta\right)$$

Based on the definition of $\csc(\theta)$ and $\sin(\theta)$, we can write:

$$\csc(\theta) = \frac{1}{\sin(\theta)} = \frac{1}{\cos\left(\widehat{COB}\right)} = \frac{1}{\frac{Adjacent\ for\ \widehat{COB}}{Hypotenuse\ for\ \widehat{COB}}} = \frac{1}{\frac{OC}{OB}} = \frac{1}{\frac{1}{OB}} = OB$$

Choice (2) is the answer.

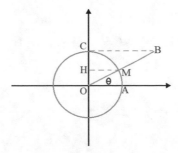

Figure 2.5 The graph of the solution of problem 2.27

2.28. From trigonometry, we know that:

$$\tan(\alpha + \beta) = \frac{\tan(\alpha) + \tan(\beta)}{1 - \tan(\alpha)\tan(\beta)}$$

$$\text{arc}\left(\tan\left(\frac{2}{3}\right)\right) \triangleq \alpha \Longrightarrow \tan(\alpha) = \frac{2}{3}$$

$$\text{arc}\left(\tan\left(\frac{1}{5}\right)\right) \triangleq \beta \Longrightarrow \tan(\beta) = \frac{1}{5}$$

Therefore,

$$\tan(\alpha + \beta) = \frac{\tan(\alpha) + \tan(\beta)}{1 - \tan(\alpha)\tan(\beta)} = \frac{\frac{2}{3} + \frac{1}{5}}{1 - \frac{2}{15}} = \frac{\frac{13}{15}}{\frac{13}{15}} = 1 \Longrightarrow \alpha + \beta = \text{arc}(\tan(1)) = \frac{\pi}{4}$$

Choice (2) is the answer.

2.29. From trigonometry, we know that:

$$\text{arc}(\tan(\alpha)) + \text{arc}\left(\tan\left(\frac{1}{\alpha}\right)\right) = \begin{cases} \dfrac{\pi}{2} & \text{if } \alpha > 0 \\[2mm] -\dfrac{\pi}{2} & \text{if } \alpha < 0 \end{cases}$$

$$\text{arc}(\cot(\alpha)) + \text{arc}(\cot(-\alpha)) = \pi$$

Therefore,

$$\text{arc}(\tan(m)) + \text{arc}\left(\tan\left(\frac{1}{m}\right)\right) + \text{arc}(\cot(m)) + \text{arc}(\cot(-m)) = \pi + \begin{cases} \dfrac{\pi}{2} & \text{if } m > 0 \\[2mm] -\dfrac{\pi}{2} & \text{if } m < 0 \end{cases}$$

$$= \begin{cases} \dfrac{3\pi}{2} & \text{if } m > 0 \\[2mm] \dfrac{\pi}{2} & \text{if } m < 0 \end{cases}$$

Choice (2) is the answer.

2.30. From trigonometry, we know that:

$$\cos(\alpha - \beta) = \cos(\alpha)\cos(\beta) + \sin(\alpha)\sin(\beta)$$

Therefore,

$$-1 \leq \cos(4x)\cos(2x) + \sin(4x)\sin(2x) \leq 0 \Longrightarrow -1 \leq \cos(4x - 2x) \leq 0 \Longrightarrow -1 \leq \cos(2x) \leq 0$$

Since x is an acute angle:

$$\Longrightarrow \frac{\pi}{2} \leq 2x \leq \pi \Longrightarrow \frac{\pi}{4} \leq x \leq \frac{\pi}{2}$$

Choice (4) is the answer.

2.31. From trigonometry, we know that:

$$\tan(\alpha - \beta) = \frac{\tan(\alpha) - \tan(\beta)}{1 + \tan(\alpha)\tan(\beta)}$$

Therefore,

$$\tan(2y) = \tan((x+y) - (x-y)) = \frac{\tan(x+y) - \tan(x-y)}{1 + \tan(x+y)\tan(x-y)} = \frac{5-7}{1 + 5 \times 7} = \frac{-2}{1 + 35} = \frac{-1}{18}$$

Choice (2) is the answer.

2.32. From trigonometry, we know that:

$$\sin(\alpha) + \cos(\alpha) = \sqrt{2}\sin\left(\alpha + \frac{\pi}{4}\right)$$

$$\sin(\alpha) - \cos(\alpha) = \sqrt{2}\sin\left(\alpha - \frac{\pi}{4}\right)$$

Therefore,

$$\frac{\sin\left(\frac{5\pi}{12}\right) + \cos\left(\frac{5\pi}{12}\right)}{\sin\left(\frac{5\pi}{12}\right) - \cos\left(\frac{5\pi}{12}\right)} = \frac{\sqrt{2}\sin\left(\frac{5\pi}{12} + \frac{\pi}{4}\right)}{\sqrt{2}\sin\left(\frac{5\pi}{12} - \frac{\pi}{4}\right)} = \frac{\sin\left(\frac{2\pi}{3}\right)}{\sin\left(\frac{\pi}{6}\right)} = \frac{\frac{\sqrt{3}}{2}}{\frac{1}{2}} = \sqrt{3}$$

Choice (1) is the answer.

2.33. From trigonometry, we know that:

$$y = a\sin(mx) \Rightarrow T = \frac{2\pi}{|m|}$$

Therefore,

$$\Rightarrow 6 = \frac{2\pi}{|b\pi|} \Rightarrow |b| = \frac{1}{3} \Rightarrow b = \pm\frac{1}{3}$$

Based on the graph and the function, the positive value of b is acceptable.

$$\Rightarrow b = \frac{1}{3}$$

Moreover, based on $y = a\sin(b\pi x)$ and the given graph, it is concluded that $a = 2$. Therefore,

$$\Rightarrow a + b = 2 + \frac{1}{3} = \frac{7}{3}$$

Choice (3) is the answer.

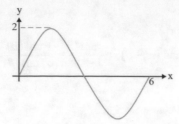

Figure 2.6 The graph of the solution of problem 2.33

2.34. From trigonometry, we know that:

$$y = a\sin(mx) \Rightarrow T = \frac{2\pi}{|m|}$$

Therefore,

$$\Rightarrow 3 = 3 \times \frac{2\pi}{|b\pi|} = 1 \Rightarrow |b| = 2 \Rightarrow b = \pm 2$$

Based on the graph and the given function, the negative value of b is accepted.

$$\Rightarrow b = -2$$

In addition, based on $y = a\sin(b\pi x)$ and the given graph, it is clear that $a = 3$. Therefore,

$$\Rightarrow a \times b = 3 \times (-2) = -6$$

Choice (1) is the answer.

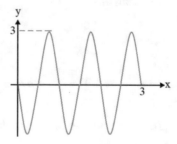

Figure 2.7 The graph of the solution of problem 2.34

2.35. From trigonometry, we know that:

$$y = a\sin\left(\frac{\pi}{2} + b\pi x\right) = a\cos(b\pi x)$$

$$y = a\cos(mx) \Rightarrow T = \frac{2\pi}{|m|}$$

Therefore,

$$\Rightarrow 3.5 - (-2.5) = 3 \times \frac{2\pi}{|b\pi|} \Rightarrow 6 = \frac{6}{|b|} \Rightarrow |b| = 1 \Rightarrow b = \pm 1$$

Based on the graph and $y = a \cos(b\pi x)$, the positive value of b is accepted.

$$\Rightarrow b = 1$$

In addition, based on $y = a \cos(b\pi x)$ and the given graph, it is clear that $a = 2$. Therefore,

$$\Rightarrow a \times b = 2 \times 1 = 2$$

Choice (1) is the answer.

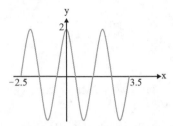

Figure 2.8 The graph of the solution of problem 2.35

2.36. From trigonometry, we know that:

$$\cos(\alpha) = \sin\left(\frac{\pi}{2} - \alpha\right)$$

$$\sin(\alpha) = \cos\left(\frac{\pi}{2} - \alpha\right)$$

$$\sin(2\alpha) = 2\sin(\alpha)\cos(\alpha)$$

Therefore,

$$\frac{\cos(5°)\cos(10°)\cos(20°)}{\cos(50°)} = \frac{\cos(5°)\cos(10°)\cos(20°)}{\sin(40°)} = \frac{\cos(5°)\cos(10°)\cos(20°)}{2\sin(20°)\cos(20°)}$$

$$= \frac{\cos(5°)\cos(10°)}{2 \times 2\sin(10°)\cos(10°)} = \frac{\cos(5°)}{4 \times 2\sin(5°)\cos(5°)} = \frac{1}{8\sin(5°)} = \frac{1}{8\cos(85°)}$$

Choice (2) is the answer.

2.37. From trigonometry, we know that:

$$\sin(x) = \frac{2\tan\left(\frac{x}{2}\right)}{1 + \tan^2\left(\frac{x}{2}\right)}$$

$$\cos(x) = \frac{1 - \tan^2\left(\frac{x}{2}\right)}{1 + \tan^2\left(\frac{x}{2}\right)}$$

Therefore,

$$\sin(x) + \cos(x) = \frac{7}{5} \Rightarrow \frac{2\tan\left(\frac{x}{2}\right)}{1 + \tan^2\left(\frac{x}{2}\right)} + \frac{1 - \tan^2\left(\frac{x}{2}\right)}{1 + \tan^2\left(\frac{x}{2}\right)} = \frac{7}{5} \Rightarrow \frac{2\tan\left(\frac{x}{2}\right) + 1 - \tan^2\left(\frac{x}{2}\right)}{1 + \tan^2\left(\frac{x}{2}\right)} = \frac{7}{5}$$

$$\Rightarrow 12\tan^2\left(\frac{x}{2}\right) - 10\tan\left(\frac{x}{2}\right) + 2 = 0 \Rightarrow \tan\left(\frac{x}{2}\right) = \frac{10 \pm \sqrt{10^2 - 4 \times 12 \times 2}}{24} = \frac{10 \pm 2}{24}$$

$$\Rightarrow \tan\left(\frac{x}{2}\right) = \frac{1}{2} \ or \ \frac{1}{3}$$

Choice (2) is the answer.

2.38. From trigonometry, we know that:

$$\sin(2\alpha) = 2\sin(\alpha)\cos(\alpha)$$

$$\sin^2(\alpha) + \cos^2(\alpha) = 1$$

$$\cos^2(\alpha) - \sin^2(\alpha) = \cos(2\alpha)$$

$$\cot(2\alpha) = \frac{\cos(2\alpha)}{\sin(2\alpha)}$$

In addition, from the factoring rule, we know that:

$$a^4 - b^4 = \left(a^2 - b^2\right)\left(a^2 + b^2\right)$$

Therefore,

$$\frac{\sin^4(\alpha) - \cos^4(\alpha)}{\sin(\alpha)\cos(\alpha)} = \frac{\left(\sin^2(\alpha) - \cos^2(\alpha)\right)\left(\sin^2(\alpha) + \cos^2(\alpha)\right)}{\sin(\alpha)\cos(\alpha)} = \frac{-\cos(2\alpha) \times 1}{\frac{1}{2}\sin(2\alpha)} = -2\cot(2\alpha)$$

Choice (2) is the answer.

2.39. Based on the information given in the problem, we have:

$$\sin^4(\alpha) + \cos^4(\alpha) = \frac{1}{2}$$

From trigonometry, we know that:

$$\sin(2\alpha) = 2\sin(\alpha)\cos(\alpha)$$

$$\sin^2(\alpha) + \sin^2(\alpha) = 1 \Rightarrow \left(\sin^2(\alpha) + \sin^2(\alpha)\right)^2 = 1$$

$$\Rightarrow \sin^4(\alpha) + \cos^4(\alpha) + 2\sin^2(\alpha)\cos^2(\alpha) = 1$$

Therefore,

$$\frac{1}{2} + 2\sin^2(\alpha)\cos^2(\alpha) = 1 \Rightarrow 4\sin^2(\alpha)\cos^2(\alpha) = 1 \Rightarrow (2\sin(\alpha)\cos(\alpha))^2 = 1$$

$$\Rightarrow \sin^2(2\alpha) = \pm 1 \Rightarrow \cos^2(2\alpha) = 0 \Rightarrow \frac{\cos^2(2\alpha)}{\sin^2(2\alpha)} = 0 \Rightarrow \cot^2(2\alpha) = 0$$

Choice (3) is the answer.

2.40. From trigonometry, we know that:

$$\sin(2x) = 2\sin(x)\cos(x)$$

$$\cos(2x) = \cos^2(x) - \sin^2(x)$$

Therefore,

$$\sin^3(x)\cos(x) - \cos^3(x)\sin(x) + 3\sin^2(x)\cos^2(x)$$

$$= \sin(x)\cos(x)\left(\sin^2(x) - \cos^2(x)\right) + \frac{3}{4} \times 4\sin^2(x)\cos^2(x)$$

$$= \frac{1}{2}\sin(2x)(-\cos(2x)) + \frac{3}{4}\sin^2(2x) = -\frac{1}{4}\sin(4x) + \frac{3}{4}\sin^2(2x)$$

$$x = \frac{3\pi}{8} \Rightarrow -\frac{1}{4}\sin\left(4 \times \frac{3\pi}{8}\right) + \frac{3}{4}\sin^2\left(2 \times \frac{3\pi}{8}\right) = -\frac{1}{4}(-1) + \frac{3}{4}\left(\frac{\sqrt{2}}{2}\right)^2 = \frac{1}{4} + \frac{3}{8} = \frac{5}{8}$$

Choice (2) is the answer.

2.41. From trigonometry, we know that:

$$\sin(\alpha) + \sin(\beta) = 2\sin\left(\frac{\alpha+\beta}{2}\right)\cos\left(\frac{\alpha-\beta}{2}\right)$$

$$\cos(\alpha) + \cos(\beta) = 2\cos\left(\frac{\alpha+\beta}{2}\right)\cos\left(\frac{\alpha-\beta}{2}\right)$$

Therefore,

$$\frac{\sin(2\alpha) + \sin(5\alpha) + \sin(8\alpha)}{\cos(2\alpha) + \cos(5\alpha) + \cos(8\alpha)} = \frac{\sin(8\alpha) + \sin(2\alpha) + \sin(5\alpha)}{\cos(8\alpha) + \cos(2\alpha) + \cos(5\alpha)}$$

$$= \frac{2\sin(5\alpha)\cos(3\alpha) + \sin(5\alpha)}{2\cos(5\alpha)\cos(3\alpha) + \cos(5\alpha)} = \frac{\sin(5\alpha)(2\cos(3\alpha) + 1)}{\cos(5\alpha)(2\cos(3\alpha) + 1)} = \tan(5\alpha)$$

$$\alpha = \frac{\pi}{15} \Rightarrow \tan(5\alpha) = \tan\left(\frac{\pi}{3}\right) = \sqrt{3}$$

Choice (3) is the answer.

2.42. From trigonometry, we know that:

$$\sin^2(x) + \cos^2(x) = 1$$

$$2\sin(x)\cos(x) = \sin(2x)$$

In addition, from the factoring rule, we know that:

$$(a+b)(a-b) = a^2 - b^2$$

Therefore,

$$(\sin(x) - \cos(x) + 2)(\sin(x) - \cos(x) - 2) = (\sin(x) - \cos(x))^2 - 4 = \sin^2(x) + \cos^2(x)$$

$$-2\sin(x)\cos(x) - 4 = 1 - \sin(2x) - 4 = -3 - \sin(2x)$$

$$x = \frac{\pi}{12} \Rightarrow -3 - \sin\left(\frac{\pi}{6}\right) = -3 - \frac{1}{2} = -\frac{7}{2}$$

Choice (4) is the answer.

2.43. From trigonometry, we know that:

$$\tan(\alpha)\cot(\alpha) = 1$$

$$1 + \tan^2(\alpha) = \frac{1}{\cos^2(\alpha)}$$

$$1 + \cot^2(\alpha) = \frac{1}{\sin^2(\alpha)}$$

$$\sin^2(\alpha) + \cos^2(\alpha) = 1$$

Therefore,

$$4\sin^2(\alpha)\cos^2(\alpha)(\tan(\alpha) + \cot(\alpha))^2$$

$$= 4\sin^2(\alpha)\cos^2(\alpha)(\tan^2(\alpha) + \cot^2(\alpha) + 2\tan(\alpha)\cot(\alpha))$$

$$= 4\sin^2(\alpha)\cos^2(\alpha)(1 + \tan^2(\alpha) + 1 + \cot^2(\alpha)) = 4\sin^2(\alpha)\cos^2(\alpha)\left(\frac{1}{\cos^2(\alpha)} + \frac{1}{\sin^2(\alpha)}\right)$$

$$= 4\sin^2(\alpha)\cos^2(\alpha)\left(\frac{\sin^2(\alpha) + \cos^2(\alpha)}{\sin^2(\alpha)\cos^2(\alpha)}\right) = 4\sin^2(\alpha)\cos^2(\alpha)\left(\frac{1}{\sin^2(\alpha)\cos^2(\alpha)}\right) = 4$$

Choice (4) is the answer.

2.44. From trigonometry, we know the common solution of the equations below.

$$\sin(\alpha) = 0 \Rightarrow \alpha = k\pi, \forall k \in \mathbb{Z}$$

$$\cos(\alpha) = 0 \Rightarrow \alpha = k\pi + \frac{\pi}{2}, \forall k \in \mathbb{Z}$$

$$\tan(\alpha) = -1 \Rightarrow \alpha = k\pi - \frac{\pi}{4}, \forall k \in \mathbb{Z}$$

Hence,

$$\sin(\pi x)\cos^2(\pi x) + \sin^2(\pi x)\cos(\pi x) = 0 \Rightarrow \sin(\pi x)\cos(\pi x)(\cos(\pi x) + \sin(\pi x)) = 0$$

$$\Rightarrow \begin{cases} \sin(\pi x) = 0 \Rightarrow \pi x = k\pi \Rightarrow x = k \xrightarrow{-2 \le x \le 2} x = -2, -1, 0, 1, 2 \\[2mm] \cos(\pi x) = 0 \Rightarrow \pi x = k\pi + \frac{\pi}{2} \Rightarrow x = k + \frac{1}{2} \xrightarrow{-2 \le x \le 2} x = -\frac{3}{2}, -\frac{1}{2}, \frac{1}{2}, \frac{3}{2} \\[2mm] \sin(\pi x) + \cos(\pi x) = 0 \Rightarrow \tan(\pi x) = -1 \Rightarrow \pi x = k\pi - \frac{\pi}{4} \Rightarrow x = k - \frac{1}{4} \xrightarrow{-2 \le x \le 2} x = -\frac{5}{4}, -\frac{1}{4}, \frac{3}{4}, \frac{7}{4} \end{cases}$$

Therefore, the number of roots of the equation is $5 + 4 + 4 = 13$. Choice (3) is the answer.

2.45. From trigonometry, we know that:

$$y = a + b\cos(mx) \Rightarrow T = \frac{2\pi}{|m|}$$

Therefore,

$$\Rightarrow 4\pi = \frac{2\pi}{|m|} \Rightarrow |m| = \frac{1}{2} \Rightarrow m = \pm\frac{1}{2}$$

Based on the graph and the function, the positive value of m is accepted.

$$\Rightarrow m = \frac{1}{2} \Rightarrow y = \frac{1}{2} + 2\cos\left(\frac{1}{2}x\right)$$

The value of function for $x = \frac{16\pi}{3}$ is:

$$y\left(\frac{16\pi}{3}\right) = \frac{1}{2} + 2\cos\left(\frac{1}{2} \times \frac{16\pi}{3}\right) = \frac{1}{2} + 2\cos\left(\frac{8\pi}{3}\right) = \frac{1}{2} + 2\cos\left(2\pi + \frac{2\pi}{3}\right) = \frac{1}{2} + 2\cos\left(\frac{2\pi}{3}\right)$$

$$= \frac{1}{2} + 2\cos\left(\pi - \frac{\pi}{3}\right) = \frac{1}{2} - 2\cos\left(\frac{\pi}{3}\right) = \frac{1}{2} - 2\left(\frac{1}{2}\right) = -\frac{1}{2}$$

Choice (1) is the answer.

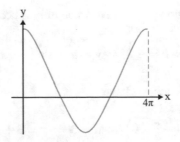

Figure 2.9   The graph of the solution of problem 2.45

2.46. From trigonometry, we know that:

$$y = a + b \sin (mx) \Rightarrow T = \frac{2\pi}{|m|}$$

Therefore,

$$\Rightarrow \frac{2\pi}{3} = \frac{2\pi}{|m|} \Rightarrow |m| = 3 \Rightarrow m = \pm 3$$

Based on the graph and the given function, the positive value of m is accepted.

$$\Rightarrow m = 3 \Rightarrow y = 1 - \sin (3x)$$

The value of function for $x = \frac{7\pi}{6}$ is:

$$y\left(\frac{7\pi}{6}\right) = 1 - \sin\left(3 \times \frac{7\pi}{6}\right) = 1 - \sin\left(\frac{7\pi}{2}\right) = 1 - \sin\left(2\pi + \frac{3\pi}{2}\right) = 1 - \sin\left(\frac{3\pi}{2}\right) = 1 - (-1) = 2$$

Choice (4) is the answer.

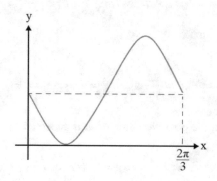

Figure 2.10 The graph of the solution of problem 2.46

2.47. From trigonometry, we know that:

$$y = a + b \sin (mx) \Rightarrow T = \frac{2\pi}{|m|}$$

Therefore,

$$\Rightarrow 5 - 1 = \frac{2\pi}{|b\pi|} \Rightarrow b = \pm \frac{1}{2}$$

Based on the graph and the given function, the positive value of m is accepted.

$$\Rightarrow b = -\frac{1}{2} \Rightarrow y = a + \sin\left(-\frac{\pi}{2}x\right)$$

By testing the point of (0, 3) in the function, we have:

$$3 = a + \sin\left(-\frac{\pi}{2} \times 0\right) \Rightarrow a = 3 \Rightarrow y = 3 + \sin\left(-\frac{\pi}{2}x\right)$$

The value of function for $x = \frac{25}{3}$ is:

$$y\left(\frac{25}{3}\right) = 3 + \sin\left(-\frac{\pi}{2} \times \frac{25}{3}\right) = 3 + \sin\left(-4\pi - \frac{\pi}{6}\right)$$

$$= 3 + \sin\left(-\frac{\pi}{6}\right) = 3 - \frac{1}{2} = 2.5$$

Choice (2) is the answer.

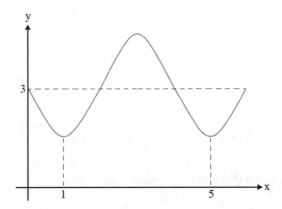

Figure 2.11 The graph of the solution of problem 2.47

2.48. By testing the point of (0, 0) in the function, we have:

$$0 = a + b\cos\left(\frac{\pi}{2} \times 0\right) \Longrightarrow a + b = 0 \qquad (1)$$

Based on the function and the graph given in the problem, we can write:

$$y_{max} = a + |b| \Longrightarrow a + |b| = 4 \qquad (2)$$

The assumption of $b < 0$ is not acceptable because it results in the equations with an impossible solution, as can be seen in the following.

$$\xrightarrow{\;Using\;(1),(2)\;} \begin{cases} a + b = 0 \\ a + b = 4 \end{cases} \Rightarrow \text{Impossible}$$

However, for the assumption of $b > 0$, we have:

$$\xrightarrow{\;Using\;(1),(2)\;} \begin{cases} a + b = 0 \\ a - b = 4 \end{cases} \Rightarrow 2b = -4 \Longrightarrow b = -2$$

Choice (1) is the answer.

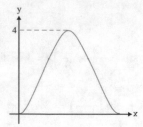

Figure 2.12 The graph of the solution of problem 2.48

2.49. From trigonometry, we know that:

$$y = 1 + a\sin(mx) \Longrightarrow T = \frac{2\pi}{|m|}$$

Therefore,

$$\Longrightarrow \frac{4}{3} = 2 \times \frac{2\pi}{|b\pi|} \Longrightarrow |b| = 3 \Longrightarrow b = \pm 3$$

Based on the function and the graph given in the problem, we can write:

$$y_{min} = 1 - |a| \Longrightarrow -1 = 1 - |a| \Longrightarrow |a| = 2 \Longrightarrow a = \pm 2$$

Based on the graph and the given function, both a and b must be either positive or negative. Hence:

$$\begin{cases} a = 2 \\ b = 3 \end{cases} \Longrightarrow a + b = 5$$

$$\begin{cases} a = -2 \\ b = -3 \end{cases} \Longrightarrow a + b = -5$$

Only $a + b = 5$ exists in the choices. Choice (3) is the answer.

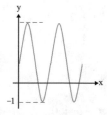

Figure 2.13 The graph of the solution of problem 2.49

2.50. From trigonometry, we know that:

$$\cos\left(\alpha + \frac{\pi}{2}\right) = -\sin(\alpha)$$

Therefore,

$$y = a - 2\cos\left(bx + \frac{\pi}{2}\right) = a + 2\sin(bx)$$

In addition, from trigonometry, we know that:

$$y = a + 2\sin(bx) \Rightarrow T = \frac{2\pi}{|b|} \Rightarrow \frac{13\pi}{18} - \frac{\pi}{18} = \frac{2\pi}{|b|} \Rightarrow |b| = 3 \Rightarrow b = \pm3$$

Based on the graph and the simplified function, i.e., $y = a + 2\sin(bx)$, the positive value of b is acceptable.

$$\Rightarrow b = 3$$

Based on the simplified function and the given graph, we can write:

$$y_{max} = a + 2 \Rightarrow 1 = a + 2 \Rightarrow a = -1$$

Hence,

$$a + b = -1 + 3 = 2$$

Choice (4) is the answer.

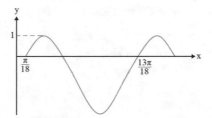

Figure 2.14 The graph of the solution of problem 2.50

2.51. From trigonometry, we know that:

$$\tan(45°) = 1$$

$$\tan(\alpha - \beta) = \frac{\tan(\alpha) - \tan(\beta)}{1 + \tan(\alpha)\tan(\beta)}$$

$$\cot(\alpha) = \frac{1}{\tan(\alpha)}$$

In addition, based on the information given in the problem, we have:

$$\tan(\alpha + 20°) = \frac{3}{4}$$

Therefore,

$$\cot\left(25^\circ - \alpha\right) = \frac{1}{\tan\left(25^\circ - \alpha\right)} = \frac{1}{\tan\left(45^\circ - \left(\alpha + 20^\circ\right)\right)} = \frac{1 + \tan\left(45^\circ\right)\tan\left(\alpha + 20^\circ\right)}{\tan\left(45^\circ\right) - \tan\left(\alpha + 20^\circ\right)}$$

$$= \frac{1 + \tan\left(\alpha + 20^\circ\right)}{1 - \tan\left(\alpha + 20^\circ\right)} = \frac{1 + \frac{3}{4}}{1 - \frac{3}{4}} = 7$$

Choice (3) is the answer.

2.52. From trigonometry, we know that:

$$\cos\left(\alpha\right) = -\sqrt{1 - \sin^2\left(\alpha\right)} \ \text{ for an obtuse angle}$$

$$\tan\left(\alpha\right) = \frac{\sin\left(\alpha\right)}{\cos\left(\alpha\right)}$$

$$\tan\left(\frac{\pi}{4}\right) = 1$$

$$\tan\left(\alpha + \beta\right) = \frac{\tan\left(\alpha\right) + \tan\left(\beta\right)}{1 - \tan\left(\alpha\right)\tan\left(\beta\right)}$$

In addition, based on the information given in the problem, we have:

$$\sin\left(\alpha\right) = \frac{3}{5}$$

Therefore,

$$\cos\left(\alpha\right) = -\sqrt{1 - \sin^2\left(\alpha\right)} = -\sqrt{1 - \left(\frac{3}{5}\right)^2} = -\frac{4}{5}$$

$$\tan\left(\alpha\right) = \frac{\sin\left(\alpha\right)}{\cos\left(\alpha\right)} = \frac{\frac{3}{5}}{-\frac{4}{5}} = -\frac{3}{4}$$

$$\tan\left(\frac{\pi}{4} + \alpha\right) = \frac{\tan\left(\frac{\pi}{4}\right) + \tan\left(a\right)}{1 - \tan\left(\frac{\pi}{4}\right)\tan\left(a\right)} = \frac{1 + \tan\left(\alpha\right)}{1 - \tan\left(\alpha\right)} = \frac{1 + \left(-\frac{3}{4}\right)}{1 - \left(-\frac{3}{4}\right)} = \frac{\frac{1}{4}}{\frac{7}{4}} = \frac{1}{7}$$

Choice (3) is the answer.

2.53. From trigonometry, we know that:

$$\tan\left(\frac{\pi}{2} - \alpha\right) = \cot\left(\alpha\right)$$

$$\cot\left(\alpha\right) = \frac{1}{\tan\left(\alpha\right)}$$

$$\tan\left(\frac{\pi}{4}\right) = 1$$

$$\tan(\alpha - \beta) = \frac{\tan(\alpha) - \tan(\beta)}{1 + \tan(\alpha)\tan(\beta)}$$

In addition, based on the information given in the problem, we have:

$$\tan\left(\frac{\pi}{2} - \alpha\right) = \frac{2}{3}$$

Therefore,

$$\frac{2}{3} = \tan\left(\frac{\pi}{2} - \alpha\right) = \cot(\alpha) = \frac{1}{\tan(\alpha)} \implies \tan(\alpha) = \frac{3}{2}$$

$$\tan\left(\frac{\pi}{4} - \alpha\right) = \frac{\tan\left(\frac{\pi}{4}\right) - \tan(a)}{1 + \tan\left(\frac{\pi}{4}\right)\tan(a)} = \frac{1 - \tan(\alpha)}{1 + \tan(\alpha)} = \frac{1 - \frac{3}{2}}{1 + \frac{3}{2}} = \frac{-\frac{1}{2}}{\frac{5}{2}} = -\frac{1}{5}$$

Choice (2) is the answer.

2.54. From trigonometry, we know that:

$$\tan(\alpha + \beta) = \frac{\tan(\alpha) + \tan(\beta)}{1 - \tan(\alpha)\tan(\beta)}$$

In addition, based on the information given in the problem, we have:

$$\tan(a + b) = \frac{2}{5}$$

$$\tan(a - b) = \frac{3}{7}$$

Therefore,

$$\tan(2a) = \tan((a + b) + (a - b)) = \frac{\tan(a + b) + \tan(a - b)}{1 - \tan(a + b)\tan(a - b)} = \frac{\frac{2}{5} + \frac{3}{7}}{1 - \frac{2}{5} \times \frac{3}{7}} = \frac{\frac{29}{35}}{\frac{29}{35}} = 1$$

Choice (4) is the answer.

2.55. From trigonometry, we know that:

$$\sin(\alpha) = \cos\left(\frac{\pi}{2} - \alpha\right)$$

$$\cos(\alpha) = \cos(-\alpha)$$

$$\tan\left(\frac{\pi}{4}\right) = 1$$

$$\tan(\alpha - \beta) = \frac{\tan(\alpha) - \tan(\beta)}{1 + \tan(\alpha)\tan(\beta)}$$

Therefore,

$$\sin\left(\frac{\pi}{4}+x\right) = \cos\left(\frac{\pi}{2}-\left(\frac{\pi}{4}-x\right)\right) = \cos\left(\frac{\pi}{4}-x\right) = \cos\left(x-\frac{\pi}{4}\right)$$

$$\Rightarrow 2 = \frac{\sin\left(x-\frac{\pi}{4}\right)}{\sin\left(x+\frac{\pi}{4}\right)} = \frac{\sin\left(x-\frac{\pi}{4}\right)}{\cos\left(x-\frac{\pi}{4}\right)} \Rightarrow \tan\left(x-\frac{\pi}{4}\right) = 2 \Rightarrow \frac{\tan\left(x\right)-\tan\left(\frac{\pi}{4}\right)}{1+\tan\left(x\right)\tan\left(\frac{\pi}{4}\right)} = 2$$

$$\Rightarrow \frac{\tan\left(x\right)-1}{1+\tan\left(x\right)} = 2 \Rightarrow \tan\left(x\right)-1 = 2+2\tan\left(x\right) \Rightarrow \tan\left(x\right) = -3$$

Choice (1) is the answer.

2.56. From trigonometry, we know that:

$$\tan\left(\frac{\pi}{4}\right) = 1$$

$$\tan\left(\alpha+\beta\right) = \frac{\tan\left(\alpha\right)+\tan\left(\beta\right)}{1-\tan\left(\alpha\right)\tan\left(\beta\right)}$$

Moreover, based on the information given in the problem, we have:

$$\alpha+\beta = \frac{\pi}{4}$$

If we calculate the tangent value of each side of the abovementioned relation, we will have:

$$\tan\left(\alpha+\beta\right) = \tan\left(\frac{\pi}{4}\right) \Rightarrow \frac{\tan\left(\alpha\right)+\tan\left(\beta\right)}{1-\tan\left(\alpha\right)\tan\left(\beta\right)} = 1 \Rightarrow \tan\left(\alpha\right)+\tan\left(\beta\right) = 1-\tan\left(\alpha\right)\tan\left(\beta\right)$$

Therefore,

$$\left(1+\tan\left(\alpha\right)\right)\left(1+\tan\left(\beta\right)\right) = 1+\tan\left(\alpha\right)+\tan\left(\beta\right)+\tan\left(\alpha\right)\tan\left(\beta\right)$$

$$= 1+\left(1-\tan\left(\alpha\right)\tan\left(\beta\right)\right)+\tan\left(\alpha\right)\tan\left(\beta\right) = 2$$

Choice (2) is the answer.

2.57. From trigonometry, we know that:

$$\tan\left(\frac{\pi}{4}\right) = 1$$

$$\tan\left(\alpha+\beta\right) = \frac{\tan\left(\alpha\right)+\tan\left(\beta\right)}{1-\tan\left(\alpha\right)\tan\left(\beta\right)}$$

$$\tan\left(\alpha-\beta\right) = \frac{\tan\left(\alpha\right)-\tan\left(\beta\right)}{1+\tan\left(\alpha\right)\tan\left(\beta\right)}$$

$$\tan\left(2\alpha\right) = \frac{2\tan\left(\alpha\right)}{1-\tan^{2}\left(\alpha\right)}$$

Therefore,

$$\tan\left(\frac{\pi}{4}+\alpha\right)-\tan\left(\frac{\pi}{4}-\alpha\right)=\frac{\tan\left(\frac{\pi}{4}\right)+\tan(\alpha)}{1-\tan\left(\frac{\pi}{4}\right)\tan(\alpha)}-\frac{\tan\left(\frac{\pi}{4}\right)-\tan(\alpha)}{1+\tan\left(\frac{\pi}{4}\right)\tan(\alpha)}$$

$$=\frac{1+\tan(\alpha)}{1-\tan(\alpha)}-\frac{1-\tan(\alpha)}{1+\tan(\alpha)}=\frac{(1+\tan(\alpha))^2-(1-\tan(\alpha))^2}{1-\tan^2(\alpha)}=\frac{4\tan(\alpha)}{1-\tan^2(\alpha)}=2\tan(2\alpha)$$

Choice (1) is the answer.

2.58. From trigonometry, we know that:

$$\tan\left(\frac{\pi}{4}\right)=1$$

$$\tan(\alpha-\beta)=\frac{\tan(\alpha)-\tan(\beta)}{1+\tan(\alpha)\tan(\beta)}$$

$$\tan(2\alpha)=\frac{2\tan(\alpha)}{1-\tan^2(\alpha)}$$

Moreover, based on the information given in the problem, we have:

$$\tan\left(\frac{\pi}{4}-\alpha\right)=\frac{1}{5}\Longrightarrow\frac{\tan\left(\frac{\pi}{4}\right)-\tan(\alpha)}{1+\tan\left(\frac{\pi}{4}\right)\tan(\alpha)}=\frac{1-\tan(\alpha)}{1+\tan(\alpha)}=\frac{1}{5}\Rightarrow5-5\tan(\alpha)=1+\tan(\alpha)\Rightarrow\tan(\alpha)=\frac{2}{3}$$

$$\tan(2\alpha)=\frac{2\tan(\alpha)}{1-\tan^2(\alpha)}=\frac{2\times\frac{2}{3}}{1-\left(\frac{2}{3}\right)^2}=\frac{12}{5}=2.4$$

Choice (3) is the answer.

2.59. From trigonometry, we know that:

$$\tan(2\alpha)=\frac{2\tan(\alpha)}{1-\tan^2(\alpha)}$$

$$\tan(\alpha-\beta)=\frac{\tan(\alpha)-\tan(\beta)}{1+\tan(\alpha)\tan(\beta)}$$

Moreover, based on the information given in the problem, we have:

$$\tan(\alpha)=2$$

$$\tan(\beta)=\frac{1}{3}$$

Therefore,

$$\tan(2\alpha)=\frac{2\tan(\alpha)}{1-\tan^2(\alpha)}\Longrightarrow\tan(2\alpha)=\frac{2\times2}{1-2^2}=-\frac{4}{3}$$

$$\tan(2\alpha - \beta) = \frac{\tan(2\alpha) - \tan(\beta)}{1 + \tan(2\alpha)\tan(\beta)} \Rightarrow \tan(2\alpha - \beta) = \frac{-\frac{4}{3} - \frac{1}{3}}{1 + \left(-\frac{4}{3}\right)\left(\frac{1}{3}\right)} = \frac{-\frac{5}{3}}{\frac{5}{9}} = -3$$

Choice (1) is the answer.

2.60. From trigonometry, we know that:

$$\cos(\pi - x) = -\cos(x)$$

$$\cos(\alpha) = \cos(\alpha_0) \Rightarrow \alpha = 2k\pi \pm \alpha_0, \forall k \in \mathbb{Z}$$

Moreover, based on the information given in the problem, we have:

$$\cos(x) \neq 0$$

Therefore,

$$\cos(3x) + \cos(x) = 0 \Rightarrow \cos(3x) = -\cos(x) \Rightarrow \cos(3x) = \cos(\pi - x)$$

$$\Rightarrow 3x = 2k\pi \pm (\pi - x) \Rightarrow \begin{cases} 3x = 2k\pi + \pi - x \Rightarrow 4x = 2k\pi + \pi \\ 3x = 2k\pi - \pi + x \Rightarrow 2x = 2k\pi - \pi \end{cases} \Rightarrow \begin{cases} x = \frac{k\pi}{2} + \frac{\pi}{4} \\ x = k\pi - \frac{\pi}{2} \end{cases}$$

However,

$$\cos(x) \neq 0 \Rightarrow x = \frac{k\pi}{2} + \frac{\pi}{4}$$

Choice (1) is the answer.

2.61. From trigonometry, we know that:

$$\cot(\alpha) = \tan\left(\frac{\pi}{2} - \alpha\right)$$

$$\tan(\alpha) = \tan(\alpha_0) \Rightarrow \alpha = k\pi + \alpha_0, \forall k \in \mathbb{Z}$$

Therefore,

$$\tan(4x) = \cot(x) \Rightarrow \tan(4x) = \tan\left(\frac{\pi}{2} - x\right)$$

$$\Rightarrow 4x = k\pi + \left(\frac{\pi}{2} - x\right) \Rightarrow 5x = k\pi + \frac{\pi}{2} \Rightarrow x = \frac{k\pi}{5} + \frac{\pi}{10}$$

$$\Rightarrow \begin{cases} k = -1 \Rightarrow x_1 = -\frac{\pi}{10} \text{ is not a positive angle} \\ k = 0 \Rightarrow x_2 = \frac{\pi}{10} \text{ is an acute angle} \\ k = 1 \Rightarrow x_3 = \frac{3\pi}{10} \text{ is an acute angle} \\ k = 2 \Rightarrow x_4 = \frac{\pi}{2} \text{ is not an acute angle} \end{cases} \Rightarrow x_2 + x_3 = \frac{2\pi}{5}$$

Choice (1) is the answer.

2.62. From trigonometry, we know that:

$$\sin^2(x) + \cos^2(x) = 1$$

$$\cos(x) = \cos(x_0) \Rightarrow x = 2k\pi \pm x_0$$

Therefore,

$$2\sin^2(x) + 3\cos(x) = 0 \Rightarrow 2(1 - \cos^2(x)) + 3\cos(x) = 0 \Rightarrow 2\cos^2(x) - 3\cos(x) - 2 = 0$$

$$\Rightarrow \cos^2(x) - \frac{3}{2}\cos(x) - 1 = 0 \Rightarrow \left(\cos(x) + \frac{1}{2}\right)(\cos(x) - 2) = 0$$

$$\Rightarrow \begin{cases} \cos(x) = -\frac{1}{2} \Rightarrow x = 2k\pi \pm \frac{2\pi}{3} \\ \cos(x) = 2 \Rightarrow \text{not acceptable} \end{cases}$$

Choice (1) is the answer.

2.63. From trigonometry, we know that:

$$\sin^2(x) + \cos^2(x) = 1$$

$$\cos(x) = \cos(x_0) \Rightarrow x = 2k\pi \pm x_0$$

Therefore,

$$2\sin^2(x) = 3\cos(x) \Rightarrow 2(1 - \cos^2(x)) - 3\cos(x) = 0 \Rightarrow 2\cos^2(x) + 3\cos(x) - 2 = 0$$

$$\Rightarrow \cos^2(x) + \frac{3}{2}\cos(x) - 1 = 0 \Rightarrow \left(\cos(x) - \frac{1}{2}\right)(\cos(x) + 2) = 0$$

$$\Rightarrow \begin{cases} \cos(x) = \frac{1}{2} \Rightarrow x = 2k\pi \pm \frac{\pi}{3} \\ \cos(x) = -2 \Rightarrow \text{not acceptable} \end{cases}$$

Choice (4) is the answer.

2.64. From trigonometry, we know that:

$$\tan(\alpha) . \cot(\alpha) = 1$$

Now, let us find the intersection point of the lines, as follows.

$$\begin{cases} x\tan(\alpha) - y\cot(\alpha) = 1 \\ x\tan(\alpha) + y\cot(\alpha) = 2 \end{cases} \Rightarrow \begin{cases} 2x\tan(\alpha) = 3 \Rightarrow x = \dfrac{3}{2\tan(\alpha)} \\ 2y\cot(\alpha) = 1 \Rightarrow y = \dfrac{1}{2\cot(\alpha)} \end{cases}$$

$$\Rightarrow xy = \frac{3}{2\tan(\alpha)} \times \frac{1}{2\cot(\alpha)} = \frac{3}{4} \Rightarrow y = \frac{3}{4x}$$

Choice (4) is the answer.

2.65. From trigonometry, we know that:

$$\sin^2(\alpha) + \cos^2(\alpha) = 1$$

Based on the information given in the problem, we have:

$$\begin{cases} x = 2 - 3\sin(\alpha) \Rightarrow \sin(\alpha) = \dfrac{x-2}{-3} \\ y = 1 + 4\cos(\alpha) \Rightarrow \cos(\alpha) = \dfrac{y-1}{4} \end{cases}$$

Therefore,

$$\Rightarrow \frac{(x-2)^2}{9} + \frac{(y-1)^2}{16} = 1$$

which is the equation of an ellipse. Choice (2) is the answer.

2.66. Based on the information given in the problem, we have:

$$\begin{cases} x = 2 - 5\cos(\alpha) \\ y = 4 \end{cases}$$

From trigonometry, we know that:

$$-1 \le \cos(\alpha) \le 1 \Rightarrow -1 \le \frac{2-x}{5} \le 1 \Rightarrow -5 \le 2 - x \le 5 \Rightarrow -7 \le -x \le 3 \Rightarrow -3 \le x \le 7$$

Therefore,

$$\Rightarrow \begin{cases} -3 \le x \le 7 \\ y = 4 \end{cases}$$

which is the equation of a horizontal line segment. Choice (3) is the answer.

2.67. From trigonometry, we know that the maximum value of cos(.) and sin(.) is one. Therefore, the only solution of the given equation is:

$$\begin{cases} \cos(x-y) = 1 \\ \sin(x+y) = 1 \end{cases}$$

The common solution of the equations can be calculated as follows:

$$\Rightarrow \begin{cases} x - y = 2k\pi \\ x + y = 2k\pi + \dfrac{\pi}{2} \end{cases} \xrightarrow{0 < x, y < 2\pi} \begin{cases} x - y = 0 \\ x + y = \dfrac{\pi}{2} \ or \ \dfrac{5\pi}{2} \end{cases} \Rightarrow y = \frac{\pi}{4} \ or \ \frac{5\pi}{4}$$

Choice (2) is the answer.

2.68. From trigonometry, we know that:

$$\tan(\alpha + \beta) = \frac{\tan(\alpha) + \tan(\beta)}{1 - \tan(\alpha)\tan(\beta)}$$

In addition, we know that the sum and the product of the roots of a quadratic equation in the form of $ax^2 + bx + c = 0$ are $-\frac{b}{a}$ and $\frac{c}{a}$, respectively.

Based on the information given in the problem, we have:

$$\tan^2(x) + (m+2)\tan(x) + 2m - 2 = 0$$

$$\alpha + \beta = \frac{\pi}{4}$$

Therefore,

$$\xrightarrow{\tan(.)} \tan(\alpha + \beta) = 1 \Rightarrow \frac{\tan(\alpha) + \tan(\beta)}{1 - \tan(\alpha)\tan(\beta)} = 1$$

$$\Rightarrow \frac{\text{sum of the roots of the quadratic equation}}{1 - \text{product of the roots of the quadratic equation}} = \frac{\frac{-(m+2)}{1}}{1 - \frac{2m-2}{1}} = \frac{-m-2}{3-2m} = 1$$

$$\Rightarrow -m - 2 = 3 - 2m \Rightarrow m = 5$$

Choice (3) is the answer.

2.69. From trigonometry, we know that:

$$\sin^6(\alpha) + \cos^6(\alpha) = 1 - 3\sin^2(\alpha)\cos^2(\alpha)$$

$$\sin^4(\alpha) + \cos^4(\alpha) = 1 - 2\sin^2(\alpha)\cos^2(\alpha)$$

Therefore,

$$\frac{\sin^6(\alpha) + \cos^6(\alpha) + 3\sin^2(\alpha)\cos^2(\alpha)}{\sin^4(\alpha) + \cos^4(\alpha) + 2\sin^2(\alpha)\cos^2(\alpha)} = \frac{1 - 3\sin^2(\alpha)\cos^2(\alpha) + 3\sin^2(\alpha)\cos^2(\alpha)}{1 - 2\sin^2(\alpha)\cos^2(\alpha) + 2\sin^2(\alpha)\cos^2(\alpha)} = 1$$

Choice (4) is the answer.

2.70. From trigonometry, we know that:

$$\sin(135^\circ) = \sin(180^\circ - 45^\circ) = \sin(45^\circ)$$

$$\cos(210^\circ) = \cos(180^\circ + 30^\circ) = -\cos(30^\circ)$$

$$\cos(135^\circ) = \cos(180^\circ - 45^\circ) = -\cos(45^\circ)$$

$$\sin(420^\circ) = \sin(360^\circ + 60^\circ) = \sin(60^\circ)$$

$$\tan(210^\circ) = \tan(180^\circ + 30^\circ) = \tan(30^\circ)$$

$$\cot\left(420^\circ\right) = \cot\left(360^\circ + 60^\circ\right) = \cot\left(60^\circ\right)$$

$$\cot\left(120^\circ\right) = \cot\left(180^\circ - 60^\circ\right) = -\cot\left(60^\circ\right)$$

$$\tan\left(330^\circ\right) = \tan\left(360^\circ - 30^\circ\right) = -\tan\left(30^\circ\right)$$

Therefore,

$$\frac{\sin\left(45^\circ\right)\left(-\cos\left(30^\circ\right)\right) + \left(-\cos\left(45^\circ\right)\right)\sin\left(60^\circ\right)}{\tan\left(30^\circ\right)\cot\left(60^\circ\right) + \left(-\cot\left(60^\circ\right)\right)\left(-\tan\left(30^\circ\right)\right)} = \frac{\frac{\sqrt{2}}{2} \times \left(\frac{-\sqrt{3}}{2}\right) + \left(-\frac{\sqrt{2}}{2}\right)\frac{\sqrt{3}}{2}}{\frac{\sqrt{3}}{3} \times \frac{\sqrt{3}}{3} + \left(-\frac{\sqrt{3}}{3}\right)\left(-\frac{\sqrt{3}}{3}\right)} = \frac{-3\sqrt{6}}{4}$$

Choice (2) is the answer.

2.71. From trigonometry, we know that:

$$\cot\left(x + y\right) = \frac{\cot\left(x\right)\cot\left(y\right) - 1}{\cot\left(x\right) + \cot\left(y\right)}$$

Based on the information given in the problem, we have:

$$x + y = k\pi + \frac{\pi}{4} \xRightarrow{k\,=\,0} x + y = \frac{\pi}{4} \xRightarrow{\cot\,(.)} \frac{\cot\left(x\right)\cot\left(y\right) - 1}{\cot\left(x\right) + \cot\left(y\right)} = 1$$

$$\Rightarrow 1 + \cot\left(x\right) + \cot\left(y\right) = \cot\left(x\right)\cot\left(y\right) \tag{1}$$

On the other hand, we can write:

$$\left(1 + \cot\left(x\right)\right)\left(1 + \cot\left(y\right)\right) = \left(1 + \cot\left(x\right) + \cot\left(y\right)\right) + \cot\left(x\right)\cot\left(y\right) \tag{2}$$

Solving (1) and (2):

$$\left(1 + \cot\left(x\right)\right)\left(1 + \cot\left(y\right)\right) = \cot\left(x\right)\cot\left(y\right) + \cot\left(x\right)\cot\left(y\right) = 2\cot\left(x\right)\cot\left(y\right)$$

Choice (4) is the answer.

2.72. From trigonometry, we know that:

$$\tan\left(\frac{3\pi}{2} - x\right) = \cot\left(x\right)$$

$$\cos\left(\frac{4\pi}{3}\right) = -\cos\left(\frac{\pi}{3}\right) = -\frac{1}{2}$$

$$\tan\left(x\right)\cot\left(x\right) = 1$$

$$\cot\left(x\right) = \frac{\cos\left(x\right)}{\sin\left(x\right)}$$

$$\cos\left(x\right) = \cos\left(x_0\right) \Rightarrow x = 2k\pi \pm x_0$$

Therefore,

$$(\sin(x) - \tan(x)) \tan\left(\frac{3\pi}{2} - x\right) = \cos\left(\frac{4\pi}{3}\right) \Rightarrow (\sin(x) - \tan(x)) \cot(x) = -\frac{1}{2}$$

$$\Rightarrow \sin(x)\cot(x) - \tan(x)\cot(x) = -\frac{1}{2} \Rightarrow \cos(x) - 1 = -\frac{1}{2} \Rightarrow \cos(x) = \frac{1}{2} \Rightarrow x = 2k\pi \pm \frac{\pi}{3}$$

Choice (3) is the answer.

2.73. From trigonometry, we know that:

$$\sin(\alpha + \beta) = \sin(\alpha)\cos(\beta) + \sin(\alpha)\cos(\beta)$$

$$\cos(\alpha + \beta) = \cos(\alpha)\cos(\beta) - \sin(\alpha)\sin(\beta)$$

$$\tan(x) = \frac{\sin(x)}{\cos(x)}$$

$$\tan(x) = \tan(x_0) \Rightarrow x = k\pi + x_0$$

Therefore,

$$\sin(2x)(\sin(x) + \cos(x)) = \cos(2x)(\cos(x) - \sin(x))$$

$$\Rightarrow \sin(2x)\sin(x) + \sin(2x)\cos(x) = \cos(2x)\cos(x) - \cos(2x)\sin(x)$$

$$\Rightarrow \sin(2x)\cos(x) + \cos(2x)\sin(x) = \cos(2x)\cos(x) - \sin(2x)\sin(x)$$

$$\Rightarrow \sin(2x + x) = \cos(2x + x) \Rightarrow \sin(3x) = \cos(3x) \xrightarrow[]{\times \dfrac{1}{\cos(3x)} \quad \& \cos(3x) \neq 0} \tan(3x) = 1$$

$$\Rightarrow 3x = k\pi + \frac{\pi}{4} \Rightarrow x = \frac{k\pi}{3} + \frac{\pi}{12} \xrightarrow{k = 0, 1, 2 \,\& x \in [0, \pi]} x_1 = \frac{\pi}{12}, x_2 = \frac{5\pi}{12}, x_3 = \frac{9\pi}{12}$$

$$\Rightarrow x_1 + x_2 + x_3 = \frac{5\pi}{4}$$

Choice (2) is the answer.

2.74. From trigonometry, we know that:

$$\sin\left(\frac{5\pi}{2} + x\right) = \sin\left(\frac{\pi}{2} + x\right) = \cos(x)$$

$$\sin(\alpha + \beta) = \sin(\alpha)\cos(\beta) + \sin(\alpha)\cos(\beta)$$

$$\sin\left(\frac{\pi}{4}\right) = \cos\left(\frac{\pi}{4}\right) = \frac{\sqrt{2}}{2}$$

$$\sin(x) = \sin(x_0) \Rightarrow \begin{cases} x = 2k\pi + x_0 \\ x = 2k\pi + \pi - x_0 \end{cases}$$

Therefore,

$$\Rightarrow \sqrt{2}\sin\left(\frac{\pi}{4} - x\right) = 1 + \cos(x) \Rightarrow \sqrt{2}\left(\sin\left(\frac{\pi}{4}\right)\cos(x) - \cos\left(\frac{\pi}{4}\right)\sin(x)\right) = 1 + \cos(x)$$

$$\Rightarrow \cos(x) - \sin(x) = 1 + \cos(x) \Rightarrow \sin(x) = -1 \Rightarrow \begin{cases} x = 2k\pi + \left(-\frac{\pi}{2}\right) \\ x = 2k\pi + \pi - \left(-\frac{\pi}{2}\right) \end{cases}$$

$$\Rightarrow \begin{cases} x = 2k\pi - \frac{\pi}{2} \\ x = 2k\pi + \frac{3\pi}{2} \end{cases} \Rightarrow x = 2k\pi - \frac{\pi}{2}$$

Choice (3) is the answer.

2.75. From trigonometry, we know that:

$$\tan\left(\frac{\pi}{3}\right) = \sqrt{3}$$

$$\tan(x) = \frac{\sin(x)}{\cos(x)}$$

$$\cos\left(\frac{\pi}{3}\right) = \frac{1}{2}$$

$$\cos(\alpha - \beta) = \cos(\alpha)\cos(\beta) + \sin(\alpha)\sin(\beta)$$

$$\cos(x) = \cos(x_0) \Rightarrow x = 2k\pi \pm x_0$$

Therefore,

$$\cos(2x) + \sqrt{3}\sin(2x) = 1 \Rightarrow \cos(2x) + \tan\left(\frac{\pi}{3}\right)\sin(2x) = 1 \Rightarrow \cos(2x) + \frac{\sin\left(\frac{\pi}{3}\right)}{\cos\left(\frac{\pi}{3}\right)}\sin(2x) = 1$$

$$\Rightarrow \cos(2x)\cos\left(\frac{\pi}{3}\right) + \sin\left(\frac{\pi}{3}\right)\sin(2x) = \cos\left(\frac{\pi}{3}\right) \Rightarrow \cos\left(2x - \frac{\pi}{3}\right) = \cos\left(\frac{\pi}{3}\right)$$

$$\Rightarrow 2x - \frac{\pi}{3} = 2k\pi \pm \frac{\pi}{3} \Rightarrow \begin{cases} 2x - \frac{\pi}{3} = 2k\pi + \frac{\pi}{3} \Rightarrow x = k\pi + \frac{\pi}{3} \\ 2x - \frac{\pi}{3} = 2k\pi - \frac{\pi}{3} \Rightarrow x = k\pi \end{cases}$$

Choice (4) is the answer.

Reference

1. Rahmani-Andebili, M. (2020). Precalculus: Practice problems, methods, and solutions, Springer Nature, 2020.

Problems: Limits

Abstract

In this chapter, the basic and advanced problems of limits are presented. The subjects include limits by direct substitution, limits by factoring, limits by rationalization, limits at infinity, trigonometric limits, limits of absolute value functions, limits involving Euler's number, limits by L'Hopital's rule, application of Taylor series in limits, and limits and continuity. To help students study the chapter in the most efficient way, the problems are categorized based on their difficulty levels (easy, normal, and hard) and calculation amounts (small, normal, and large). Moreover, the problems are ordered from the easiest problem with the smallest computations to the most difficult problems with the largest calculations.

3.1. Calculate the value of the following limit [1].

$$\lim_{x \to (-1)^+} \frac{[x] + 1}{x^2 - 1}$$

Difficulty level   ● Easy   ○ Normal   ○ Hard
Calculation amount   ● Small   ○ Normal   ○ Large
1) $-\frac{1}{2}$
2) 0
3) $\frac{1}{2}$
4) ∞

3.2. Calculate the limit of the following function if $x \to 2^+$.

$$f(x) = \frac{x + 4}{[-x] - 3}$$

Difficulty level   ● Easy   ○ Normal   ○ Hard
Calculation amount   ● Small   ○ Normal   ○ Large
1) 1
2) −1
3) 2
4) −2

© Springer Nature Switzerland AG 2021
M. Rahmani-Andebili, *Calculus*, https://doi.org/10.1007/978-3-030-64980-7_3

3.3. Determine the value of the following limit.

$$\lim_{x \to 0^-} \frac{x+2}{[x]}$$

Difficulty level   ● Easy   ○ Normal   ○ Hard
Calculation amount   ● Small   ○ Normal   ○ Large
1) 2
2) −2
3) 1
4) −1

3.4. Determine the value of the limit below.

$$\lim_{x \to -\infty} \frac{[x]+3x}{[x]-3x}$$

Difficulty level   ● Easy   ○ Normal   ○ Hard
Calculation amount   ● Small   ○ Normal   ○ Large
1) 2
2) −2
3) 4
4) −4

3.5. Calculate the limit of the function below if $x \to 0$.

$$f(x) = \frac{x + \sqrt[3]{x}}{x - \sqrt[3]{x}}$$

Difficulty level   ● Easy   ○ Normal   ○ Hard
Calculation amount   ● Small   ○ Normal   ○ Large
1) 0
2) 1
3) −1
4) ∞

3.6. Calculate the value of the following limit.

$$\lim_{x \to 0^-} \frac{[x]}{x}$$

Difficulty level   ● Easy   ○ Normal   ○ Hard
Calculation amount   ● Small   ○ Normal   ○ Large
1) ∞
2) −∞
3) 0
4) 1

3.7. Determine the limit of the function below if $x \to 0^+$.

$$f(x) = \frac{(x^2 - 1)\sqrt{x}}{(x\sqrt{x} + 1)x}$$

Difficulty level   ● Easy   ○ Normal   ○ Hard
Calculation amount   ● Small   ○ Normal   ○ Large
1) ∞
2) $-\infty$
3) 0
4) -1

3.8. Determine the value of the following limit.

$$\lim_{x \to 0^+} \left(\frac{1}{x} - \frac{1}{x^3} \right)$$

Difficulty level   ● Easy   ○ Normal   ○ Hard
Calculation amount   ● Small   ○ Normal   ○ Large
1) ∞
2) $-\infty$
3) 0
4) 1

3.9. For the function below, calculate the value of $\lim\limits_{x \to 1^+} f(x) - \lim\limits_{x \to 1^-} f(x)$.

$$f(x) = \frac{2x}{[2x] + 2}$$

Difficulty level   ● Easy   ○ Normal   ○ Hard
Calculation amount   ● Small   ○ Normal   ○ Large
1) $-\infty$
2) $-\frac{1}{6}$
3) $\frac{2}{3}$
4) ∞

3.10. Calculate the value of $\lim\limits_{x \to 2^+} ([x] - 2)[x]$.

Difficulty level   ● Easy   ○ Normal   ○ Hard
Calculation amount   ● Small   ○ Normal   ○ Large
1) -2
2) -1
3) 0
4) 1

3.11. Calculate the limit of the following function if $x \to 4^-$.

$$f(x) = \frac{[x] - 4}{x^2 - 16}$$

Difficulty level   ● Easy   ○ Normal   ○ Hard
Calculation amount   ● Small   ○ Normal   ○ Large

1) 0

2) $\frac{1}{8}$

3) ∞

4) $-\infty$

3.12. Determine the value of the limit below.

$$\lim_{x \to 1^-} \frac{1 - x^3}{\arc(\cos(x))}$$

Difficulty level   ○ Easy   ● Normal   ○ Hard
Calculation amount ● Small   ○ Normal   ○ Large

1) 1

2) −1

3) 0

4) −3

3.13. Calculate the value of the limit below.

$$\lim_{x \to 0} \frac{\tan(x) - \tan(3x) + \tan(2x)}{x^3}$$

Difficulty level   ○ Easy   ● Normal   ○ Hard
Calculation amount ● Small   ○ Normal   ○ Large

1) −6

2) 6

3) 10

4) −10

3.14. Calculate the value of the following limit.

$$\lim_{x \to 3} \frac{9 - x^2}{2 - \sqrt{x + 1}}$$

Difficulty level   ○ Easy   ● Normal   ○ Hard
Calculation amount ● Small   ○ Normal   ○ Large

1) 6

2) 12

3) 18

4) 24

3.15. Determine the limit of the function below if $x \to +\infty$.

$$f(x) = \frac{\sin(x)}{x}$$

Difficulty level   ○ Easy   ● Normal   ○ Hard
Calculation amount ● Small   ○ Normal   ○ Large

1) Undefined

2) 0

3) 1

4) ∞

3.16. Determine the value of the limit below.

$$\lim_{x \to 0} \frac{[x^2] - x^2}{x \tan(x)}$$

Difficulty level   ○ Easy   ● Normal   ○ Hard
Calculation amount   ● Small   ○ Normal   ○ Large
1) 1
2) −1
3) 2
4) −2

3.17. Determine the value of the following limit.

$$\lim_{x \to +\infty} x \sin\left(\frac{1}{x}\right)$$

Difficulty level   ○ Easy   ● Normal   ○ Hard
Calculation amount   ● Small   ○ Normal   ○ Large
1) 1
2) −1
3) 0
4) Undefined

3.18. Calculate the value of the following limit.

$$\lim_{x \to -\infty} \left(\frac{x^2 + x - 1}{-3x + 4\sqrt{-x}}\right)$$

Difficulty level   ○ Easy   ● Normal   ○ Hard
Calculation amount   ● Small   ○ Normal   ○ Large
1) $\frac{1}{3}$
2) $-\frac{1}{3}$
3) ∞
4) $-\infty$

3.19. Calculate the value of the limit below.

$$\lim_{x \to 0^+} \frac{(x + 1)\sqrt{x}}{x^2 - x}$$

Difficulty level   ○ Easy   ● Normal   ○ Hard
Calculation amount   ● Small   ○ Normal   ○ Large
1) 0
2) −1
3) ∞
4) $-\infty$

3.20. Determine the limit of the following function if $x \rightarrow +\infty$.

$$f(x) = \frac{x}{x - 1 + \sqrt{x^2 + x - 1}}$$

Difficulty level   ○ Easy   ● Normal   ○ Hard
Calculation amount   ● Small   ○ Normal   ○ Large
1) ∞
2) 0
3) $-\frac{1}{2}$
4) $\frac{1}{2}$

3.21. Calculate the value of $\lim\limits_{x \to 0} x \cot(x)$.

Difficulty level   ○ Easy   ● Normal   ○ Hard
Calculation amount   ● Small   ○ Normal   ○ Large
1) 0
2) ∞
3) 1
4) 2

3.22. For what value of "a", the following function has a definite limit at $x = 1$?

$$f(x) = \begin{cases} x^2 + ax & x > 1 \\ x - 3 & x < 1 \end{cases}$$

Difficulty level   ○ Easy   ● Normal   ○ Hard
Calculation amount   ● Small   ○ Normal   ○ Large
1) 0
2) 3
3) -3
4) -2

3.23. Determine the value of the limit below.

$$\lim_{x \to 2^-} \frac{|x^3 - 8|}{x - \sqrt{2x}}$$

Difficulty level   ○ Easy   ● Normal   ○ Hard
Calculation amount   ● Small   ○ Normal   ○ Large
1) -24
2) -16
3) 16
4) 24

3.24. Calculate the value of the limit below.

$$\lim_{x \to 0^-} \frac{[x] + x}{[-x] + x}$$

Difficulty level   ○ Easy   ● Normal   ○ Hard
Calculation amount   ● Small   ○ Normal   ○ Large

1) $+\infty$
2) $-\infty$
3) 1
4) -1

3.25. Determine the value of the limit below.

$$\lim_{x \to 0} \frac{\sin(3x) + \sin(7x)}{3x + \tan(2x)}$$

Difficulty level   ○ Easy   ● Normal   ○ Hard
Calculation amount   ● Small   ○ Normal   ○ Large
1) 1
2) 2
3) -1
4) -2

3.26. Calculate the value of the following limit.

$$\lim_{x \to 0} \frac{\sqrt{x+3} - \sqrt{3}}{x}$$

Difficulty level   ○ Easy   ● Normal   ○ Hard
Calculation amount   ● Small   ○ Normal   ○ Large
1) $\frac{\sqrt{3}}{3}$
2) $\frac{\sqrt{3}}{6}$
3) $\frac{\sqrt{3}}{2}$
4) $\frac{\sqrt{3}}{9}$

3.27. Calculate the value of the following limit.

$$\lim_{x \to 0} \frac{1 - \cos(x)}{\sin(x)}$$

Difficulty level   ○ Easy   ● Normal   ○ Hard
Calculation amount   ● Small   ○ Normal   ○ Large
1) 0
2) 1
3) -1
4) $\sqrt{2}$

3.28. Calculate the value of the limit below.

$$\lim_{x \to 0} \frac{5x - \sin(x)}{2x + \cos(x) - 1}$$

Difficulty level   ○ Easy   ● Normal   ○ Hard
Calculation amount   ● Small   ○ Normal   ○ Large

1) 1
2) 2
3) −1
4) −2

3.29. Determine the value of the limit below.

$$\lim_{x \to 2^-} \frac{x^3 - 8}{|x - 2|} + 5x$$

Difficulty level   ○ Easy   ● Normal   ○ Hard
Calculation amount   ● Small   ○ Normal   ○ Large
1) 2
2) −2
3) 1
4) −1

3.30. Calculate the value of the limit below.

$$\lim_{x \to \left(\frac{\pi}{2}\right)^+} \frac{\sin(x) + \cos(x)}{\cos(x)}$$

Difficulty level   ○ Easy   ● Normal   ○ Hard
Calculation amount   ● Small   ○ Normal   ○ Large
1) ∞
2) −∞
3) 0
4) 1

3.31. Calculate the value of the following limit.

$$\lim_{x \to 0} \frac{3x^4 + 2x^3}{(\arc(\sin(x)))^3}$$

Difficulty level   ○ Easy   ● Normal   ○ Hard
Calculation amount   ● Small   ○ Normal   ○ Large
1) 1
2) 2
3) 0
4) ∞

3.32. Determine the value of the limit below.

$$\lim_{x \to -\infty} \left[\frac{2}{x + 1}\right] x$$

Difficulty level   ○ Easy   ● Normal   ○ Hard
Calculation amount   ● Small   ○ Normal   ○ Large
1) ∞
2) 2
3) 0
4) −∞

3.33. Calculate the value of the following limit.

$$\lim_{x \to 3^+} \frac{x-4}{\sqrt{x^2 - 4x + 3}}$$

Difficulty level   ○ Easy   ● Normal   ○ Hard
Calculation amount   ● Small   ○ Normal   ○ Large
1) ∞
2) $-\infty$
3) 1
4) -1

3.34. Calculate the value of $\lim\limits_{x \to -\infty} \left(x + \sqrt{x^2 + 4x - 10} \right)$.

Difficulty level   ○ Easy   ● Normal   ○ Hard
Calculation amount   ● Small   ○ Normal   ○ Large
1) 2
2) -2
3) ∞
4) $-\infty$

3.35. Determine the value of the limit below.

$$\lim_{x \to 2} \frac{4 - x^2}{6 - 2\sqrt{x^2 + 5}}$$

Difficulty level   ○ Easy   ● Normal   ○ Hard
Calculation amount   ● Small   ○ Normal   ○ Large
1) 0
2) 2
3) 3
4) 1

3.36. Calculate the value of the following limit.

$$\lim_{x \to 0} \frac{\sin (2x)}{\sqrt{x+1} - 1}$$

Difficulty level   ○ Easy   ● Normal   ○ Hard
Calculation amount   ● Small   ○ Normal   ○ Large
1) 2
2) 4
3) 3
4) 1

3.37. Calculate the limit of $\sqrt{x^4 + 2x^2 + x} - x^2$ if $x \to -\infty$.

Difficulty level   ○ Easy   ● Normal   ○ Hard
Calculation amount   ○ Small   ● Normal   ○ Large
1) 1
2) $+\infty$
3) 0
4) $-\infty$

3.38. Calculate the value of the limit below.

$$\lim_{x \to -3} \frac{\left|x^2 - 9\right|}{x + 3}$$

Difficulty level   ○ Easy   ● Normal   ○ Hard
Calculation amount   ○ Small   ● Normal   ○ Large
1) 6
2) −6
3) 3
4) Undefined

3.39. Calculate the value of the following limit.

$$\lim_{x \to \frac{1}{2}} \frac{\tan \frac{\pi x}{2} - 1}{\cos (\pi x)}$$

Difficulty level   ○ Easy   ● Normal   ○ Hard
Calculation amount   ○ Small   ● Normal   ○ Large
1) 1
2) −1
3) 2
4) −2

3.40. Calculate the limit of $\sqrt{x + 5} - \sqrt{x + 1}$ if $x \to \infty$.

Difficulty level   ○ Easy   ● Normal   ○ Hard
Calculation amount   ○ Small   ● Normal   ○ Large
1) 4
2) 2
3) 0
4) ∞

3.41. Calculate the value of the following limit.

$$\lim_{x \to \frac{\pi}{2}} \frac{\tan (2x) \cos (x)}{1 + \cos (2x)}$$

Difficulty level   ○ Easy   ● Normal   ○ Hard
Calculation amount   ○ Small   ● Normal   ○ Large
1) 1
2) $\frac{1}{2}$
3) −1
4) $-\frac{1}{2}$

3.42. Determine the value of the following limit.

$$\lim_{x \to 0^-} \frac{\tan (2x)}{\sqrt{1 - \cos (x)}}$$

Difficulty level   ○ Easy   ● Normal   ○ Hard
Calculation amount   ○ Small   ● Normal   ○ Large

1) $-2\sqrt{2}$

2) $-\sqrt{2}$

3) $\sqrt{2}$

4) $2\sqrt{2}$

3.43. Calculate the value of $\lim\limits_{x\to-\infty}\left(\sqrt[3]{n+1000}-\sqrt[3]{n-20}\right)$.

Difficulty level   ○ Easy   ● Normal   ○ Hard
Calculation amount   ○ Small   ● Normal   ○ Large

1) 2

2) 0

3) 10

4) 20

3.44. Calculate the value of the limit below.

$$\lim_{x\to\pi^+}\frac{\sin\left(\pi\sin\left(x\right)\right)\sin\left(\frac{x}{2}\right)}{\sqrt{1+\cos\left(x\right)}}$$

Difficulty level   ○ Easy   ● Normal   ○ Hard
Calculation amount   ○ Small   ● Normal   ○ Large

1) $-\pi\sqrt{2}$

2) -2π

3) π^2

4) $\pi^2\sqrt{2}$

3.45. Calculate the value of the following limit.

$$\lim_{x\to0}\frac{\sqrt[3]{1+x^2}-\sqrt[4]{1-2x}}{2x^2+2x}$$

Difficulty level   ○ Easy   ● Normal   ○ Hard
Calculation amount   ○ Small   ● Normal   ○ Large

1) $\frac{1}{4}$

2) $-\frac{1}{4}$

3) 1

4) -1

3.46. Calculate the limit of the function below if $x\to0$.

$$f(x)=\frac{\sin^2(x)+\sin(x)+\cos^2(x)-\cos(x)}{\sin^2(x)-\sin(x)+\cos^2(x)-\cos(x)}$$

Difficulty level   ○ Easy   ● Normal   ○ Hard
Calculation amount   ○ Small   ● Normal   ○ Large

1) 1

2) -1

3) 2

4) -2

3.47. Determine the value of the following limit.

$$\lim_{x \to 0} \frac{\cos(mx) - \cos(nx)}{x^2}$$

Difficulty level   ○ Easy   ● Normal   ○ Hard
Calculation amount   ○ Small   ● Normal   ○ Large

1) $n^2 + m^2$
2) $n^2 - m^2$
3) $\frac{n^2 - m^2}{2}$
4) $\frac{n^2 + m^2}{2}$

3.48. Calculate the value of the limit below.

$$\lim_{x \to 0} \frac{\sin(x) - x}{\tan(x) - x}$$

Difficulty level   ○ Easy   ● Normal   ○ Hard
Calculation amount   ○ Small   ● Normal   ○ Large

1) $-\frac{1}{2}$
2) $\frac{1}{2}$
3) $-\frac{1}{4}$
4) $\frac{1}{4}$

3.49. Calculate the value of the following limit.

$$\lim_{x \to \pi} \frac{1 + \cos^3(x)}{1 - \cos^2(x)}$$

Difficulty level   ○ Easy   ● Normal   ○ Hard
Calculation amount   ○ Small   ● Normal   ○ Large

1) $\frac{3}{2}$
2) $-\frac{3}{2}$
3) 3
4) −3

3.50. Calculate the limit of the function below if $n \to +\infty$.

$$f(n) = \frac{3n^2}{\sqrt{5^n}}$$

Difficulty level   ○ Easy   ○ Normal   ● Hard
Calculation amount   ● Small   ○ Normal   ○ Large

1) 0
2) 1
3) ∞
4) $\frac{3}{\sqrt{5}}$

3.51. Determine the value of the limit below.

$$\lim_{x \to 0} \frac{x^3 - \sin(x)(1 - \cos(x))}{x^3}$$

Difficulty level   ○ Easy   ○ Normal   ● Hard
Calculation amount   ○ Small   ● Normal   ○ Large
1) 0
2) $\frac{1}{2}$
3) 1
4) −1

3.52. Calculate the value of the following limit.

$$\lim_{x \to 1^-} \frac{\arc(\cos x)}{\sqrt{1 - x}}$$

Difficulty level   ○ Easy   ○ Normal   ● Hard
Calculation amount   ○ Small   ● Normal   ○ Large
1) $\sqrt{2}$
2) $-\sqrt{2}$
3) $-\frac{\sqrt{2}}{2}$
4) $\frac{\sqrt{2}}{2}$

3.53. Determine the value of n in the following equation.

$$\lim_{x \to 1} \left(x^2 - 1\right) \cot\left(x^n - 1\right) = \frac{1}{2}$$

Difficulty level   ○ Easy   ○ Normal   ● Hard
Calculation amount   ○ Small   ● Normal   ○ Large
1) 8
2) 4
3) $\frac{1}{8}$
4) $\frac{1}{4}$

3.54. Calculate the limit of $\sin(4x)(\cot(2x) - \cot(x))$ if $x \to 0$.

Difficulty level   ○ Easy   ○ Normal   ● Hard
Calculation amount   ○ Small   ● Normal   ○ Large
1) 4
2) 2
3) −2
4) −4

3.55. Calculate the value of the following limit.

$$\lim_{x \to 0^-} \frac{\sin(x) - x}{\frac{1}{2} \sin(2x) - x \cos(x)}$$

Difficulty level   ○ Easy   ○ Normal   ● Hard
Calculation amount   ○ Small   ● Normal   ○ Large

1) 0
2) ∞
3) 1
4) $-\infty$

3.56. Calculate the limit of the following function if $x \to \frac{\pi}{4}$.

$$f(x) = \frac{1 - \sqrt[3]{\tan(x)}}{1 - 2\sin^2(x)}$$

Difficulty level ○ Easy ○ Normal ● Hard
Calculation amount ○ Small ● Normal ○ Large
1) $\frac{1}{3}$
2) 3
3) $\frac{1}{2}$
4) 2

3.57. Calculate the value of the limit below.

$$\lim_{x \to 0} \frac{1 - \cos^3(x)}{\sin(x)\tan(2x)}$$

Difficulty level ○ Easy ○ Normal ● Hard
Calculation amount ○ Small ● Normal ○ Large
1) 4
2) -4
3) $\frac{3}{4}$
4) $-\frac{3}{4}$

3.58. Determine the value of the following limit.

$$\lim_{x \to 0^+} \frac{1 - \sqrt{\cos(x)}}{1 - \cos(\sqrt{x})}$$

Difficulty level ○ Easy ○ Normal ● Hard
Calculation amount ○ Small ○ Normal ● Large
1) 0
2) 1
3) $\frac{1}{2}$
4) 2

Reference

1. Rahmani-Andebili, M. (2020). Precalculus: Practice problems, methods, and solutions, Springer Nature, 2020.

Solutions of Problems: Limits

4

Abstract

In this chapter, the problems of the third chapter are fully solved, in detail, step-by-step, and with different methods. The subjects include limits by direct substitution, limits by factoring, limits by rationalization, limits at infinity, trigonometric limits, limits of absolute value functions, limits involving Euler's number, limits by L'Hopital's rule, application of Taylor series in limits, and limits and continuity.

4.1. The problem can be solved as follows [1].

$$\lim_{x \to (-1)^+} \frac{[x] + 1}{x^2 - 1} = \frac{(-1) + 1}{1^- - 1} = \frac{0}{0^-} = 0$$

Choice (2) is the answer.

4.2. The problem can be solved as follows.

$$\lim_{x \to 2^+} \frac{x + 4}{[-x] - 3} = \frac{2 + 4}{-3 - 3} = -1$$

Choice (2) is the answer.

4.3. The problem can be solved as follows.

$$\lim_{x \to 0^-} \frac{x + 2}{[x]} = \frac{0 + 2}{-1} = -2$$

Choice (2) is the answer.

4.4. The problem can be solved as follows.

$$\lim_{x \to -\infty} \frac{[x] + 3x}{[x] - 3x} = \lim_{x \to -\infty} \frac{4x}{-2x} = \lim_{x \to -\infty} (-2) = -2$$

Choice (2) is the answer.

© Springer Nature Switzerland AG 2021
M. Rahmani-Andebili, *Calculus*, https://doi.org/10.1007/978-3-030-64980-7_4

4.5. The problem can be solved as follows.

$$\lim_{x \to 0} \frac{x + \sqrt[3]{x}}{x - \sqrt[3]{x}} = \lim_{x \to 0} \frac{\sqrt[3]{x}\left(\sqrt[3]{x^2} + 1\right)}{\sqrt[3]{x}\left(\sqrt[3]{x^2} - 1\right)} = \lim_{x \to 0} \frac{\left(\sqrt[3]{x^2} + 1\right)}{\left(\sqrt[3]{x^2} - 1\right)} = -1$$

Choice (3) is the answer.

4.6. The problem can be solved as follows.

$$\lim_{x \to 0^-} \frac{[x]}{x} = \frac{-1}{0^-} = +\infty$$

Choice (1) is the answer.

4.7. The problem can be solved as follows.

$$\lim_{x \to 0^+} \frac{(x^2 - 1)\sqrt{x}}{(x\sqrt{x} + 1)x} = \lim_{x \to 0^+} \frac{(x^2 - 1)}{(x\sqrt{x} + 1)\sqrt{x}} = \frac{-1}{0^+} = -\infty$$

Choice (2) is the answer.

4.8. The problem can be solved as follows.

$$\lim_{x \to 0^+} \left(\frac{1}{x} - \frac{1}{x^3}\right) = \lim_{x \to 0^+} \left(\frac{x^2 - 1}{x^3}\right) - \frac{-1}{0^+} = -\infty$$

Choice (2) is the answer.

4.9. The problem can be solved as follows.

$$\lim_{x \to 1^+} f(x) - \lim_{x \to 1^-} f(x) = \lim_{x \to 1^+} \frac{2x}{[2x] + 2} - \lim_{x \to 1^-} \frac{2x}{[2x] + 2} = \lim_{x \to 1^+} \frac{2x}{2 + 2} - \lim_{x \to 1^-} \frac{2x}{1 + 2}$$

$$= \lim_{x \to 1^+} \frac{x}{2} - \lim_{x \to 1^-} \frac{2}{3}x = \frac{1}{2} - \frac{2}{3} = \frac{-1}{6}$$

Choice (2) is the answer.

4.10. The problem can be solved as follows.

$$\lim_{x \to 2^+} ([x] - 2)[x] = \lim_{x \to 2^+} (2 - 2) \times 2 = \lim_{x \to 2^+} 0 = 0$$

Choice (3) is the answer.

4.11. The problem can be solved as follows.

$$\lim_{x \to 4^-} \frac{[x] - 4}{x^2 - 16} = \lim_{x \to 4^-} \frac{3 - 4}{16^- - 16} = \frac{-1}{0^-} = +\infty$$

Choice (3) is the answer.

4.12. From trigonometry and calculus, we know that:

$$arc(\cos(1^-)) = 0^+$$

$$\frac{d}{dx}(arc(\cos x)) = \frac{-1}{\sqrt{1-x^2}}$$

The problem can be solved as follows.

$$\lim_{x \to 1^-}\frac{1-x^3}{arc(\cos(x))} = \frac{0^+}{0^+} \overset{H}{\Rightarrow} \lim_{x \to 1^-}\frac{\frac{d}{dx}(1-x^3)}{\frac{d}{dx}(arc(\cos(x)))} = \lim_{x \to 1^-}\frac{-3x^2}{\frac{-1}{\sqrt{1-x^2}}} = \lim_{x \to 1^-}3x^2\sqrt{1-x^2} = 0$$

Choice (3) is the answer.

4.13. From the application of the Taylor series in limit, we know that:

$$\lim_{x \to 0}\tan(x) \equiv x + \frac{x^3}{3}$$

The problem can be solved as follows.

$$\lim_{x \to 0}\frac{\tan(x) - \tan(3x) + \tan(2x)}{x^3} \equiv \lim_{x \to 0}\frac{x + \frac{x^3}{3} - \left(3x + \frac{(3x)^3}{3}\right) + 2x + \frac{(2x)^3}{3}}{x^3} = \lim_{x \to 0}\frac{-6x^3}{x^3} = \lim_{x \to 0}(-6) = -6$$

Choice (1) is the answer.

4.14. The problem can be solved as follows.

$$\lim_{x \to 3}\frac{9-x^2}{2-\sqrt{x+1}} = \frac{0}{0} \overset{H}{\Rightarrow} \lim_{x \to 3}\frac{\frac{d}{dx}(9-x^2)}{\frac{d}{dx}(2-\sqrt{x+1})} = \lim_{x \to 3}\frac{-2x}{\frac{-1}{2\sqrt{x+1}}} = \lim_{x \to 3}4x\sqrt{x+1} = 24$$

Choice (4) is the answer.

4.15. From trigonometry, we know that:

$$-1 \leq \sin(x) \leq 1$$

The problem can be solved as follows.

$$\lim_{x \to +\infty}\frac{\sin(x)}{x} = \lim_{x \to +\infty}\sin(x) \times \frac{1}{x} = (\text{Bounded quantity}) \times 0 = 0$$

Choice (2) is the answer.

4.16. From the application of the Taylor series in limit, we know that:

$$\lim_{x \to 0}\tan(x) \equiv x$$

The problem can be solved as follows.

$$\lim_{x \to 0} \frac{[x^2] - x^2}{x \tan(x)} = \lim_{x \to 0} \frac{0 - x^2}{x \tan(x)} \equiv \lim_{x \to 0} \frac{-x^2}{x \times x} = \lim_{x \to 0}(-1) = -1$$

Choice (2) is the answer.

4.17. From the application of the Taylor series in limit, we know that:

$$\lim_{x \to +\infty} \sin\left(\frac{1}{x}\right) \equiv \frac{1}{x}$$

The problem can be solved as follows.

$$\lim_{x \to +\infty} x \sin\left(\frac{1}{x}\right) = \lim_{x \to +\infty} x \times \frac{1}{x} = \lim_{x \to +\infty} 1 = 1$$

Choice (1) is the answer.

4.18. From calculus, we know that:

$$\lim_{x \to \pm\infty} \left(a_m x^m + a_{m-1} x^{m-1} + \dots + a_2 x^2 + a_1 x + a_0\right) \equiv a_m x^m$$

or

$$\lim_{x \to \pm\infty} \left(a_m x^m + a_n x^n\right) \equiv a_m x^m \ \ if \ \ m > n$$

Therefore,

$$\lim_{x \to -\infty} \left(\frac{x^2 + x - 1}{-3x + 4\sqrt{-x}}\right) \equiv \lim_{x \to -\infty} \frac{x^2}{-3x} = \lim_{x \to -\infty} \left(-\frac{x}{3}\right) = +\infty$$

Choice (3) is the answer.

4.19. From calculus, we know that:

$$\lim_{x \to 0} \left(a_m x^m + a_{m-1} x^{m-1} + \dots + a_{m-n} x^{m-n} + a_{m-n-1} x^{m-n-1}\right) \equiv a_{m-n-1} x^{m-n-1}$$

or

$$\lim_{x \to 0} \left(a_m x^m + a_n x^n\right) \equiv a_n x^n \ \ if \ \ m > n$$

The problem can be solved as follows.

$$\lim_{x \to 0^+} \frac{(x+1)\sqrt{x}}{x^2 - x} = \lim_{x \to 0^+} \frac{x\sqrt{x} + \sqrt{x}}{x^2 - x} \equiv \lim_{x \to 0^+} \frac{\sqrt{x}}{-x} = \lim_{x \to 0^+} \frac{-1}{\sqrt{x}} = -\infty$$

Choice (4) is the answer.

4.20. From calculus, we know that:

$$\lim_{x \to +\infty} \sqrt{x^2 + ax + b} \equiv \left| x + \frac{a}{2} \right|$$

The problem can be solved as follows.

$$\lim_{x \to +\infty} \frac{x}{x - 1 + \sqrt{x^2 + x - 1}} \equiv \lim_{x \to +\infty} \frac{x}{x - 1 + \left| x + \frac{1}{2} \right|} = \lim_{x \to +\infty} \frac{x}{2x - \frac{1}{2}} \equiv \lim_{x \to +\infty} \frac{x}{2x} = \lim_{x \to +\infty} \frac{1}{2} = \frac{1}{2}$$

Choice (4) is the answer.

4.21. From trigonometry, we know that:

$$\cot(x) = \frac{1}{\tan(x)}$$

From the application of the Taylor series in limit, we know that:

$$\lim_{x \to 0} \tan(x) \equiv x$$

The problem can be solved as follows.

$$\lim_{x \to 0} x \cot(x) = \lim_{x \to 0} \frac{x}{\tan(x)} \equiv \lim_{x \to 0} \frac{x}{x} = 1$$

Choice (3) is the answer.

4.22. As we know, the limit of a function at the point of x_0 exits if:

$$\lim_{x \to x_0^-} f(x) = \lim_{x \to x_0^+} f(x) \Rightarrow \lim_{x \to 1^-} f(x) = \lim_{x \to 1^+} f(x) \tag{1}$$

Therefore,

$$\lim_{x \to 1^-} f(x) = \lim_{x \to 1^-} (x - 3) = 1 - 3 = -2 \tag{2}$$

$$\lim_{x \to 1^+} f(x) = \lim_{x \to 1^+} (x^2 + ax) = 1 + a \tag{3}$$

$$\xRightarrow{Using\ (1),(2),(3)} -2 = 1 + a \Rightarrow a = -3$$

Choice (3) is the answer.

4.23. The problem can be solved as follows.

$$\lim_{x \to 2^-} \frac{|x^3 - 8|}{x - \sqrt{2x}} = \lim_{x \to 2^-} \frac{-(x^3 - 8)}{x - \sqrt{2x}} = \frac{0^+}{0^-}$$

$$\overset{H}{\Rightarrow} \lim_{x \to 2^-} \frac{\frac{d}{dx}(-(x^3 - 8))}{\frac{d}{dx}(x - \sqrt{2x})} = \lim_{x \to 2^-} \frac{-3x^2}{1 - \frac{1}{\sqrt{2x}}} = \frac{-3 \times 2^2}{1 - \frac{1}{\sqrt{2 \times 2}}} = \frac{-12}{\frac{1}{2}} = -24$$

Choice (1) is the answer.

4.24. The problem can be solved as follows.

$$\lim_{x \to 0^-} \frac{[x] + x}{[-x] + x} = \lim_{x \to 0^-} \frac{-1 + x}{0 + x} = \frac{-1 + 0^-}{0 + 0^-} = \frac{-1}{0^-} = +\infty$$

Choice (1) is the answer.

4.25. The problem can be solved as follows.

$$\lim_{x \to 0} \frac{\sin(3x) + \sin(7x)}{3x + \tan(2x)} = \frac{0}{0}$$

$$\overset{H}{\Rightarrow} \lim_{x \to 0} \frac{\frac{d}{dx}(\sin(3x) + \sin(7x))}{\frac{d}{dx}(3x + \tan(2x))} = \lim_{x \to 0} \frac{3\cos(3x) + 7\cos(7x)}{3 + 2(1 + \tan^2(2x))} = \frac{3 + 7}{3 + 2} = 2$$

Choice (2) is the answer.

4.26. The problem can be solved as follows.

$$\lim_{x \to 0} \frac{\sqrt{x + 3} - \sqrt{3}}{x} = \lim_{x \to 0} \frac{\sqrt{x + 3} - \sqrt{3}}{x} \times \frac{\sqrt{x + 3} + \sqrt{3}}{\sqrt{x + 3} + \sqrt{3}} = \lim_{x \to 0} \frac{x + 3 - 3}{x(\sqrt{x + 3} + \sqrt{3})}$$

$$= \lim_{x \to 0} \frac{1}{(\sqrt{x + 3} + \sqrt{3})} = \frac{1}{\sqrt{3} + \sqrt{3}} = \frac{1}{2\sqrt{3}} = \frac{\sqrt{3}}{6}$$

Choice (2) is the answer.

4.27. The problem can be solved as follows.

$$\lim_{x \to 0} \frac{1 - \cos(x)}{\sin(x)} = \frac{0}{0}$$

$$\overset{H}{\Rightarrow} \lim_{x \to 0} \frac{\frac{d}{dx}(1 - \cos(x))}{\frac{d}{dx}\sin(x)} = \lim_{x \to 0} \frac{\sin(x)}{\cos(x)} = \frac{0}{1} = 0$$

Choice (1) is the answer.

4.28. The problem can be solved as follows.

$$\lim_{x \to 0} \frac{5x - \sin(x)}{2x + \cos(x) - 1} = \frac{0}{0}$$

$$\overset{H}{\Rightarrow} \lim_{x \to 0} \frac{\frac{d}{dx}(5x - \sin(x))}{\frac{d}{dx}(2x + \cos(x) - 1)} = \lim_{x \to 0} \frac{5 - \cos(x)}{2 - \sin(x)} = \frac{5 - 1}{2 - 0} = 2$$

Choice (2) is the answer.

4.29. The problem can be solved as follows.

$$\lim_{x \to 2^-} \frac{x^3 - 8}{|x - 2|} + 5x = \lim_{x \to 2^-} \left(\frac{(x - 2)(x^2 + 2x + 4)}{-(x - 2)} + 5x \right) = \lim_{x \to 2^-} \left(-x^2 - 2x - 4 + 5x \right)$$

$$\lim_{x \to 2^-} \left(-x^2 + 3x - 4 \right) = -4 + 6 - 4 = -2$$

Choice (2) is the answer.

4.30. The problem can be solved as follows.

$$\lim_{x \to \left(\frac{\pi}{2}\right)^+} \frac{\sin(x) + \cos(x)}{\cos(x)} = \frac{1^- + 0^-}{0^-} = \frac{1^-}{0^-} = -\infty$$

Choice (2) is the answer.

4.31. The problem can be solved as follows.

$$\lim_{x \to 0} \text{arc}(\sin(x)) \equiv x$$

$$\lim_{x \to 0} 3x^4 + 2x^3 \equiv 2x^3$$

$$\lim_{x \to 0} \frac{3x^4 + 2x^3}{(\text{arc}(\sin(x)))^3} \equiv \lim_{x \to 0} \frac{2x^3}{x^3} = \lim_{x \to 0} 2 = 2$$

Choice (2) is the answer.

4.32. The problem can be solved as follows.

$$\lim_{x \to -\infty} \left[\frac{2}{x + 1} \right] x = \lim_{x \to -\infty} [0^-] x = (-1)(-\infty) = +\infty$$

Choice (1) is the answer.

4.33. The problem can be solved as follows.

$$\lim_{x \to 3^+} \frac{x - 4}{\sqrt{x^2 - 4x + 3}} = \lim_{x \to 3^+} \frac{x - 4}{\sqrt{(x - 3)(x - 1)}} = \frac{-1}{0^+} = -\infty$$

Choice (2) is the answer.

4.34. From calculus, we know that:

$$\lim_{x \to \pm\infty} \sqrt{x^2 + ax + b} \equiv \left| x + \frac{a}{2} \right|$$

The problem can be solved as follows.

$$\lim_{x \to -\infty} \left(x + \sqrt{x^2 + 4x - 10} \right) \equiv \lim_{x \to -\infty} (x + |x + 2|) = \lim_{x \to -\infty} (x - x - 2) = \lim_{x \to -\infty} (-2) = -2$$

Choice (2) is the answer.

4.35. The problem can be solved as follows.

$$\lim_{x \to 2} \frac{4 - x^2}{6 - 2\sqrt{x^2 + 5}} = \frac{0}{0}$$

$$\overset{H}{\Rightarrow} \lim_{x \to 2} \frac{\frac{d}{dx}(4 - x^2)}{\frac{d}{dx}\left(6 - 2\sqrt{x^2 + 5}\right)} = \lim_{x \to 2} \frac{-2x}{-2 \times \frac{2x}{2\sqrt{x^2 + 5}}} = \lim_{x \to 2} \sqrt{x^2 + 5} = 3$$

Choice (3) is the answer.

4.36. The problem can be solved as follows.

$$\lim_{x \to 0} \frac{\sin(2x)}{\sqrt{x + 1} - 1} = \frac{0}{0}$$

$$\overset{H}{\Rightarrow} \lim_{x \to 0} \frac{\frac{d}{dx}(\sin(2x))}{\frac{d}{dx}\left(\sqrt{x + 1} - 1\right)} = \lim_{x \to 0} \frac{2\cos(2x)}{\frac{1}{2\sqrt{x + 1}}} = \frac{2 \times 1}{\frac{1}{2}} = 4$$

Choice (2) is the answer.

4.37. From calculus, we know that:

$$\lim_{x \to -\infty} \sqrt{x^4 + 2x^2 + x} \equiv x^2$$

$$\lim_{x \to \pm\infty} \left(a_m x^m + a_{m-1} x^{m-1} + \ldots + a_2 x^2 + a_1 x + a_0\right) \equiv a_m x^m$$

or,

$$\lim_{x \to \pm\infty} \left(a_m x^m + a_n x^n\right) \equiv a_m x^m \quad if \quad m > n$$

The problem can be solved as follows.

$$\lim_{x \to -\infty} \sqrt{x^4 + 2x^2 + x} - x^2 = \lim_{x \to -\infty} \left(\sqrt{x^4 + 2x^2 + x} - x^2\right) \times \frac{\left(\sqrt{x^4 + 2x^2 + x} + x^2\right)}{\left(\sqrt{x^4 + 2x^2 + x} + x^2\right)}$$

$$= \lim_{x \to -\infty} \frac{x^4 + 2x^2 + x - x^4}{\left(\sqrt{x^4 + 2x^2 + x} + x^2\right)} \equiv \lim_{x \to -\infty} \frac{2x^2 + x}{(x^2 + x^2)} = \lim_{x \to -\infty} \frac{2x^2}{2x^2} = \lim_{x \to -\infty} 1 = 1$$

Choice (1) is the answer.

4.38. From calculus, we know that the limit of a function at a specific point (x_0) exits if

$$\lim_{x \to x_0^-} f(x) = \lim_{x \to x_0^+} f(x)$$

Therefore, we must have:

$$\lim_{x \to (-3)^-} \frac{\left|x^2 - 9\right|}{x + 3} = \lim_{x \to (-3)^+} \frac{\left|x^2 - 9\right|}{x + 3} \tag{1}$$

$$\lim_{x \to (-3)^-} \frac{|x^2 - 9|}{x + 3} = \lim_{x \to (-3)^-} \frac{x^2 - 9}{x + 3} = \lim_{x \to (-3)^-} (x - 3) = -6 \qquad (2)$$

$$\lim_{x \to (-3)^+} \frac{|x^2 - 9|}{x + 3} = \lim_{x \to (-3)^+} \frac{-(x^2 - 9)}{x + 3} = \lim_{x \to (-3)^+} -(x - 3) = 6 \qquad (3)$$

$$\xRightarrow{(1), (2), (3)} -6 \neq 6 \Rightarrow \lim_{x \to -3} \frac{|x^2 - 9|}{x + 3} = \text{Undefined}$$

Choice (4) is the answer.

4.39. The problem can be solved as follows.

$$\lim_{x \to \frac{1}{2}} \frac{\tan\left(\frac{\pi x}{2}\right) - 1}{\cos(\pi x)} = \frac{0}{0}$$

$$\overset{H}{\Rightarrow} \lim_{x \to \frac{1}{2}} \frac{\frac{d}{dx}\left(\tan\left(\frac{\pi x}{2}\right) - 1\right)}{\frac{d}{dx}(\cos(\pi x))} = \lim_{x \to \frac{1}{2}} \frac{\frac{\pi}{2}\left(1 + \tan^2\left(\frac{\pi x}{2}\right)\right)}{-\pi \sin(\pi x)} = \frac{\frac{\pi}{2}(1 + 1)}{-\pi \times 1} = -1$$

Choice (2) is the answer.

4.40. The problem can be solved as follows.

$$\lim_{x \to +\infty} \left(\sqrt{x + 5} - \sqrt{x + 1}\right) = \lim_{x \to +\infty} \left(\sqrt{x + 5} - \sqrt{x + 1}\right) \times \frac{\left(\sqrt{x + 5} + \sqrt{x + 1}\right)}{\left(\sqrt{x + 5} + \sqrt{x + 1}\right)}$$

$$= \lim_{x \to +\infty} \frac{x + 5 - (x + 1)}{\left(\sqrt{x + 5} + \sqrt{x + 1}\right)} = \lim_{x \to +\infty} \frac{4}{\left(\sqrt{x + 5} + \sqrt{x + 1}\right)} = 0$$

Choice (3) is the answer.

4.41. From trigonometry, we know that:

$$1 + \cos(2x) = 2\cos^2(x)$$

$$\tan(x) = \frac{\sin(x)}{\cos(x)}$$

$$\sin(2x) = 2\sin(x)\cos(x)$$

The problem can be solved as follows.

$$\lim_{x \to \frac{\pi}{2}} \frac{\tan(2x)\cos(x)}{1 + \cos(2x)} = \lim_{x \to \frac{\pi}{2}} \frac{\sin(2x)\cos(x)}{\cos(2x) \times 2\cos^2(x)} = \lim_{x \to \frac{\pi}{2}} \frac{2\sin(x)\cos^2(x)}{\cos(2x) \times 2\cos^2(x)} = \lim_{x \to \frac{\pi}{2}} \frac{\sin(x)}{\cos(2x)} = \frac{1}{-1} = -1$$

Choice (3) is the answer.

4.42. From calculus and trigonometry, we know that:

$$1 - \cos^2(x) = \sin^2(x)$$

Moreover, from the application of the Taylor series in limit, we know that:

$$\lim_{x \to 0^-} \sin(x) \equiv x$$

$$\lim_{x \to 0^-} \tan(x) \equiv x$$

The problem can be solved as follows.

$$\lim_{x \to 0^-} \frac{\tan(2x)}{\sqrt{1 - \cos(x)}} = \lim_{x \to 0^-} \frac{\tan(2x)}{\sqrt{1 - \cos(x)}} \times \frac{\sqrt{1 + \cos(x)}}{\sqrt{1 + \cos(x)}} = \lim_{x \to 0^-} \frac{\tan(2x)\sqrt{1 + \cos(x)}}{\sqrt{1 - \cos^2(x)}}$$

$$= \lim_{x \to 0^-} \frac{\tan(2x) \times \sqrt{2}}{\sqrt{\sin^2(x)}} = \lim_{x \to 0^-} \frac{\tan(2x) \times \sqrt{2}}{|\sin(x)|} = \lim_{x \to 0^-} \frac{\tan(2x) \times \sqrt{2}}{-\sin(x)} = \lim_{x \to 0^-} \frac{2x \times \sqrt{2}}{-x}$$

$$= \lim_{x \to 0^-} \left(-2\sqrt{2}\right) = -2\sqrt{2}$$

Choice (1) is the answer.

4.43. The problem can be solved as follows.

$$\lim_{x \to -\infty} \left(\sqrt[3]{n + 1000} - \sqrt[3]{n - 20}\right)$$

$$= \lim_{x \to -\infty} \left(\sqrt[3]{n + 1000} - \sqrt[3]{n - 20}\right) \times \frac{\left(\sqrt[3]{(n + 1000)^2} + \sqrt[3]{(n + 1000)(n - 20)} + \sqrt[3]{(n - 20)^2}\right)}{\left(\sqrt[3]{(n + 1000)^2} + \sqrt[3]{(n + 1000)(n - 20)} + \sqrt[3]{(n - 20)^2}\right)}$$

$$= \lim_{x \to -\infty} \frac{n + 1000 - (n - 20)}{\sqrt[3]{(n + 1000)^2} + \sqrt[3]{(n + 1000)(n - 20)} + \sqrt[3]{(n - 20)^2}} = \frac{1020}{+\infty} = 0$$

Choice (2) is the answer.

4.44. From the application of the Taylor series in limit, we know that:

$$\lim_{x \to 0} \sin(x) \equiv x$$

In addition, from trigonometry, we know that:

$$1 + \cos(x) = 2\cos^2\left(\frac{x}{2}\right)$$

$$\sin(x) = 2\sin\left(\frac{x}{2}\right)\cos\left(\frac{x}{2}\right)$$

The problem can be solved as follows.

$$\lim_{x\to\pi^+} \frac{\sin(\pi\sin(x))\sin\left(\frac{x}{2}\right)}{\sqrt{1+\cos(x)}} \equiv \lim_{x\to\pi^+} \frac{\pi\sin(x)\sin\left(\frac{x}{2}\right)}{\sqrt{2\cos^2\left(\frac{x}{2}\right)}} = \lim_{x\to\pi^+} \frac{\pi\times 2\sin\left(\frac{x}{2}\right)\cos\left(\frac{x}{2}\right)\sin\left(\frac{x}{2}\right)}{\sqrt{2}\left|\cos\left(\frac{x}{2}\right)\right|}$$

$$= \lim_{x\to\pi^+} \frac{\pi\times 2\sin^2\left(\frac{x}{2}\right)\cos\left(\frac{x}{2}\right)}{\sqrt{2}\left(-\cos\left(\frac{x}{2}\right)\right)} = \lim_{x\to\pi^+} \left(-\pi\sqrt{2}\sin^2\left(\frac{x}{2}\right)\right) = -\pi\sqrt{2}\times 1 = -\pi\sqrt{2}$$

Choice (1) is the answer.

4.45. From the application of the Taylor series in limit, we know that:

$$\lim_{\alpha\to 0} \sqrt[n]{1+\alpha} \equiv \lim_{\alpha\to 0} 1+\frac{\alpha}{n}$$

$$\lim_{x\to 0} \left(a_m x^m + a_{m-1}x^{m-1} + \ldots + a_{m-n}x^{m-n} + a_{m-n-1}x^{m-n-1}\right) \equiv a_{m-n-1}x^{m-n-1}$$

or,

$$\lim_{x\to 0} \left(a_m x^m + a_n x^n\right) \equiv a_n x^n \ \ if \ \ m > n$$

The problem can be solved as follows.

$$\lim_{x\to 0} \frac{\sqrt[3]{1+x^2} - \sqrt[4]{1-2x}}{2x^2+2x} \equiv \lim_{x\to 0} \frac{1+\frac{x^2}{3} - \left(1-\frac{2x}{4}\right)}{2x^2+2x} = \lim_{x\to 0} \frac{\frac{x^2}{3}+\frac{x}{2}}{2x^2+2x} \equiv \lim_{x\to 0} \frac{\frac{x}{2}}{2x} = \frac{1}{4}$$

Choice (1) is the answer.

4.46. From trigonometry, we know that:

$$\sin^2(x) + \cos^2(x) = 1$$

The problem can be solved as follows.

$$\lim_{x\to 0} \frac{\sin^2(x) + \sin(x) + \cos^2(x) - \cos(x)}{\sin^2(x) - \sin(x) + \cos^2(x) - \cos(x)} = \lim_{x\to 0} \frac{1+\sin(x) - \cos(x)}{1-\sin(x) - \cos(x)} = \frac{0}{0}$$

$$\overset{H}{\Rightarrow} \lim_{x\to 0} \frac{\frac{d}{dx}\left(1+\sin(x) - \cos(x)\right)}{\frac{d}{dx}\left(1-\sin(x) - \cos(x)\right)} = \lim_{x\to 0} \frac{\cos(x)+\sin(x)}{-\cos(x)+\sin(x)} = \frac{1+0}{-1+0} = -1$$

Choice (2) is the answer.

4.47. From the application of the Taylor series in limit, we know that:

$$\lim_{u(x)\to 0} \sin(u(x)) \equiv u(x)$$

The problem can be solved as follows.

$$\lim_{x\to 0} \frac{\cos(mx) - \cos(nx)}{x^2} = \frac{0}{0}$$

$$\overset{H}{\Rightarrow} \lim_{x\to 0} \frac{\frac{d}{dx}(\cos(mx) - \cos(nx))}{\frac{d}{dx}(x^2)} = \lim_{x\to 0} \frac{-m\sin(mx) + n\sin(nx)}{2x} \equiv \lim_{x\to 0} \frac{-m(mx) + n(nx)}{2x}$$

$$= \lim_{x\to 0} \frac{-m^2 + n^2}{2} = \frac{n^2 - m^2}{2}$$

Choice (3) is the answer.

4.48. From the application of the Taylor series in limit, we know that:

$$\lim_{x\to 0} \sin(x) \equiv x$$

$$\lim_{x\to 0} \tan(x) \equiv x$$

The problem can be solved as follows.

$$\lim_{x\to 0} \frac{\sin(x) - x}{\tan(x) - x} = \frac{0}{0}$$

$$\overset{H}{\Rightarrow} \lim_{x\to 0} \frac{\frac{d}{dx}(\sin(x) - x)}{\frac{d}{dx}(\tan(x) - x)} = \lim_{x\to 0} \frac{\cos(x) - 1}{1 + \tan^2(x) - 1} = \lim_{x\to 0} \frac{\cos(x) - 1}{\tan^2(x)} = \frac{0}{0}$$

$$\overset{H}{\Rightarrow} \lim_{x\to 0} \frac{\frac{d}{dx}(\cos(x) - 1)}{\frac{d}{dx}(\tan^2(x))} = \lim_{x\to 0} \frac{-\sin(x)}{2\tan(x)(1 + \tan^2(x))} \equiv \lim_{x\to 0} \frac{-x}{2x(1 + x^2)} = \lim_{x\to 0} \frac{-1}{2(1 + x^2)}$$

$$= \frac{-1}{2(1 + 0)} = \frac{-1}{2}$$

Choice (1) is the answer.

4.49. From trigonometry, we know that:

$$1 + \cos^3(x) = (1 + \cos(x))(1 - \cos(x) + \cos^2(x))$$

$$1 - \cos^2(x) = (1 + \cos(x))(1 - \cos(x))$$

The problem can be solved as follows.

$$\lim_{x\to\pi} \frac{1 + \cos^3(x)}{1 - \cos^2(x)} = \frac{0}{0}$$

$$\Rightarrow \lim_{x\to\pi} \frac{1 + \cos^3(x)}{1 - \cos^2(x)} = \lim_{x\to\pi} \frac{(1 + \cos(x))(1 - \cos(x) + \cos^2(x))}{(1 + \cos(x))(1 - \cos(x))} = \lim_{x\to\pi} \frac{1 - \cos(x) + \cos^2(x)}{1 - \cos(x)}$$

$$= \frac{1 - (-1) + (-1)^2}{1 - (-1)} = \frac{3}{2}$$

Choice (1) is the answer.

4.50. As we know from calculus:

$$\text{If } a > 1, k \in N \implies \lim_{n \to +\infty} \frac{n^k}{a^n} = 0$$

Hence,

$$\lim_{n \to +\infty} \frac{3n^2}{\sqrt{5^n}} = \lim_{n \to +\infty} 3 \frac{n^2}{\left(\sqrt{5}\right)^n} = 3 \times 0 = 0$$

Choice (1) is the answer.

4.51. From trigonometry, we know that:

$$1 - \cos(x) = 2 \sin^2\left(\frac{x}{2}\right)$$

Moreover, from the application of the Taylor series in limit, we know that:

$$\lim_{x \to 0} \sin^n(x) \equiv x^n$$

Thus,

$$\lim_{x \to 0} \frac{x^3 - \sin(x)(1 - \cos(x))}{x^3} = \lim_{x \to 0} \frac{x^3 - \sin(x)\left(2\sin^2\left(\frac{x}{2}\right)\right)}{x^3} \equiv \lim_{x \to 0} \frac{x^3 - x \times 2\left(\frac{x}{2}\right)^2}{x^3}$$

$$= \lim_{x \to 0} \frac{x^3 - \frac{x^3}{2}}{x^3} = \lim_{x \to 0} \frac{\frac{x^3}{2}}{x^3} = \lim_{x \to 0} \frac{1}{2} = \frac{1}{2}$$

Choice (2) is the answer.

4.52. From trigonometry and calculus, we know that:

$$\text{arc}(\cos(1^-)) = 0^+$$

$$\frac{d}{dx}(\text{arc}(\cos x)) = \frac{-1}{\sqrt{1 - x^2}}$$

$$\frac{d}{dx}\left(\sqrt{1 - x}\right) = \frac{-1}{2\sqrt{1 - x}}$$

The problem can be solved as follows.

$$\lim_{x \to 1^-} \frac{\text{arc}(\cos x)}{\sqrt{1 - x}} = \frac{0^+}{0^+}$$

$$\overset{H}{\implies} \lim_{x \to 1^-} \frac{\frac{d}{dx}(\text{arc}(\cos x))}{\frac{d}{dx}\left(\sqrt{1 - x}\right)} = \lim_{x \to 1^-} \frac{\frac{-1}{\sqrt{1 - x^2}}}{\frac{-1}{2\sqrt{1 - x}}} = \lim_{x \to 1^-} \frac{\frac{1}{\sqrt{(1 - x)(1 + x)}}}{\frac{1}{2\sqrt{1 - x}}} = \lim_{x \to 1^-} \frac{2}{\sqrt{1 + x}} = \sqrt{2}$$

Choice (1) is the answer.

4.53. Based on the information given in the problem, we have:

$$\lim_{x \to 1} \left(x^2 - 1\right) \cot \left(x^n - 1\right) = \frac{1}{2} \tag{1}$$

From calculus, we know that:

$$\cot (x) = \frac{1}{\tan (x)}$$

From the application of the Taylor series in limit, we know that:

$$\lim_{u(x) \to 0} \tan \left(u(x)\right) \equiv u(x)$$

The problem can be solved as follows.

$$\lim_{x \to 1} \left(x^2 - 1\right) \cot \left(x^n - 1\right) = \lim_{x \to 1} \frac{x^2 - 1}{\tan \left(x^n - 1\right)} \equiv \lim_{x \to 1} \frac{x^2 - 1}{x^n - 1} = \frac{0}{0}$$

$$\overset{H}{\Rightarrow} \lim_{x \to 1} \frac{2x}{nx^{n-1}} = \frac{2}{n} \tag{2}$$

Solving (1) and (2):

$$\frac{2}{n} = \frac{1}{2} \Rightarrow n = 4$$

Choice (2) is the answer.

4.54. From trigonometry, we know that:

$$\sin (4x) = 2 \sin (2x) \cos (2x)$$

$$\cot (x) = \frac{\cos (x)}{\sin (x)}$$

$$\sin (x - y) = \sin (x) \cos (y) - \cos (x) \sin (y)$$

The problem can be solved as follows.

$$\lim_{x \to 0} \sin (4x)\left(\cot (2x) - \cot (x)\right) = \lim_{x \to 0} 2 \sin (2x) \cos (2x) \left(\frac{\cos (2x)}{\sin (2x)} - \frac{\cos (x)}{\sin (x)}\right)$$

$$= \lim_{x \to 0} 2 \sin (2x) \cos (2x) \left(\frac{\sin (x) \cos (2x) - \cos (x) \sin (2x)}{\sin (2x) \sin (x)}\right)$$

$$= \lim_{x \to 0} 2 \sin (2x) \cos (2x) \left(\frac{\sin (x - 2x)}{\sin (2x) \sin (x)}\right) = \lim_{x \to 0} \left(-2 \cos (2x)\right) = -2$$

Choice (3) is the answer.

4.55. From the application of the Taylor series in limit, we know that:

$$\lim_{x \to 0} \sin(x) \equiv x - \frac{x^3}{6}$$

$$\lim_{x \to 0} \cos(x) \equiv 1 - \frac{x^2}{2}$$

The problem can be solved as follows.

$$\lim_{x \to 0^-} \frac{\sin(x) - x}{\frac{1}{2}\sin(2x) - x\cos(x)} \equiv \lim_{x \to 0^-} \frac{-\frac{x^3}{6}}{\frac{1}{2}\left(2x - \frac{(2x)^3}{6}\right) - x\left(1 - \frac{x^2}{2}\right)} = \lim_{x \to 0^-} \frac{-\frac{x^3}{6}}{-\frac{x^3}{6}} = \lim_{x \to 0^-} 1 = 1$$

Choice (3) is the answer.

4.56. The problem can be solved as follows.

$$\lim_{x \to \frac{\pi}{4}} \frac{1 - \sqrt[3]{\tan(x)}}{1 - 2\sin^2(x)} = \frac{0}{0}$$

$$\overset{H}{\Rightarrow} \lim_{x \to \frac{\pi}{4}} \frac{\frac{d}{dx}\left(1 - \sqrt[3]{\tan(x)}\right)}{\frac{d}{dx}\left(1 - 2\sin^2(x)\right)} = \lim_{x \to \frac{\pi}{4}} \frac{-\frac{1 + \tan^2(x)}{3\sqrt[3]{\tan^2(x)}}}{-4\sin(x)\cos(x)} = \frac{\frac{1+1}{3 \times 1}}{4 \times \frac{\sqrt{2}}{2} \times \frac{\sqrt{2}}{2}} = \frac{1}{3}$$

Choice (1) is the answer.

4.57. From calculus and trigonometry, we know that:

$$\sin^2(x) + \cos^2(x) = 1$$

$$1 - \cos^3(x) = (1 - \cos(x))(1 + \cos(x) + \cos^2(x))$$

From the application of the Taylor series in limit, we know that:

$$\lim_{x \to 0} \sin(x) = x$$

$$\lim_{u(x) \to 0} \tan(u(x)) \equiv u(x)$$

The problem can be solved as follows.

$$\lim_{x \to 0} \frac{1 - \cos^3(x)}{\sin(x)\tan(2x)} = \frac{0}{0}$$

$$\Rightarrow \lim_{x \to 0} \frac{1 - \cos^3(x)}{\sin(x)\tan(2x)} = \lim_{x \to 0} \frac{(1 - \cos(x))(1 + \cos(x) + \cos^2(x))}{\sin(x)\tan(2x)} \times \frac{(1 - \cos(x))}{(1 + \cos(x))}$$

$$= \lim_{x \to 0} \frac{(1 - \cos^2(x))(1 + \cos(x) + \cos^2(x))}{\sin(x)\tan(2x)(1 + \cos(x))} = \lim_{x \to 0} \frac{\sin^2(x) \times (1 + 1 + 1)}{\sin(x)\tan(2x) \times (1 + 1)}$$

$$\lim_{x \to 0} \frac{3 \sin^2(x)}{2 \sin(x) \tan(2x)} \equiv \lim_{x \to 0} \frac{3x^2}{2x \times 2x} = \lim_{x \to 0} \frac{3}{4} = \frac{3}{4}$$

Choice (3) is the answer.

4.58. From trigonometry, we know that:

$$1 - \cos(x) = 2 \sin^2\left(\frac{x}{2}\right)$$

$$\sin^2(x) + \cos^2(x) = 1$$

From the application of the Taylor series in limit, we know that:

$$\lim_{u(x) \to 0^+} \sin^n(u(x)) \equiv \lim_{u(x) \to 0^+} (u(x))^n$$

The problem can be solved as follows.

$$\lim_{x \to 0^+} \frac{1 - \sqrt{\cos(x)}}{1 - \cos(\sqrt{x})} = \lim_{x \to 0^+} \frac{1 - \sqrt{\cos(x)}}{1 - \cos(\sqrt{x})} \times \frac{1 + \sqrt{\cos(x)}}{1 + \sqrt{\cos(x)}} \times \frac{1 + \cos(\sqrt{x})}{1 + \cos(\sqrt{x})}$$

$$= \lim_{x \to 0^+} \frac{(1 - \cos(x))(1 + \cos(\sqrt{x}))}{(1 - \cos^2(\sqrt{x}))(1 + \sqrt{\cos(x)})} = \lim_{x \to 0^+} \frac{(1 - \cos(x))(1 + 1)}{(1 - \cos^2(\sqrt{x}))(1 + 1)}$$

$$= \lim_{x \to 0^+} \frac{1 - \cos(x)}{1 - \cos^2(\sqrt{x})} = \lim_{x \to 0^+} \frac{2 \sin^2\left(\frac{x}{2}\right)}{\sin^2(\sqrt{x})} \equiv \lim_{x \to 0^+} \frac{2\left(\frac{x}{2}\right)^2}{(\sqrt{x})^2} = \lim_{x \to 0^+} \frac{x}{2} = 0$$

Choice (1) is the answer.

Reference

1. Rahmani-Andebili, M. (2020). Precalculus: Practice problems, methods, and solutions, Springer Nature, 2020.

Abstract

In this chapter, the basic and advanced problems of derivatives and its applications are presented. The subjects include definition of derivative, differentiation formulas, product rule, quotient rule, chain rule, derivatives of trigonometric functions, derivatives of exponential, derivatives of logarithm functions, derivatives of inverse trigonometric functions, derivatives of hyperbolic functions, implicit differentiation, higher-order derivatives, logarithmic differentiation, applications of derivatives, rates of change, critical points, minimum and maximum values, and absolute extrema. To help students study the chapter in the most efficient way, the problems are categorized based on their difficulty levels (easy, normal, and hard) and calculation amounts (small, normal, and large). Moreover, the problems are ordered from the easiest problem with the smallest computations to the most difficult problems with the largest calculations.

5.1. Calculate the value of $f'(x = 1)$ if $f(x) = xe^x - e^x$ [1].

Difficulty level   ● Easy   ○ Normal   ○ Hard
Calculation amount   ● Small   ○ Normal   ○ Large
1) 1
2) 0
3) $-e$
4) e

5.2. If $f(x) + g(x^3) = 5x - 1$ and $f'(1) = 2$, calculate the value of $g'(1)$.

Difficulty level   ● Easy   ○ Normal   ○ Hard
Calculation amount   ● Small   ○ Normal   ○ Large
1) 1
2) -1
3) 2
4) -2

5.3. Determine the range of x where the function of $y(x) = 1 - 4x^2$ is ascending.

Difficulty level   ● Easy   ○ Normal   ○ Hard
Calculation amount   ● Small   ○ Normal   ○ Large
1) $x < 0$
2) $x > 0$
3) $-2 < x < 2$
4) $-4 < x < 4$

M. Rahmani-Andebili, *Calculus*, https://doi.org/10.1007/978-3-030-64980-7_5

5.4. Determine the derivative of the function below at $x = \frac{1}{4}$.

$$f(x) = \frac{x - \sqrt{x}}{1 - \sqrt{x}}$$

Difficulty level   ● Easy   ○ Normal   ○ Hard
Calculation amount   ● Small   ○ Normal   ○ Large
1) -1
2) $-\frac{1}{2}$
3) $\frac{1}{2}$
4) 1

5.5. Determine the first derivative of the function of $(x^{100} + x^{50} + 50x^2 + 50x + 1)^{10}$ at $x = 0$.
Difficulty level   ● Easy   ○ Normal   ○ Hard
Calculation amount   ○ Small   ● Normal   ○ Large
1) 100
2) 200
3) 400
4) 500

5.6. Calculate the derivative of the function of $f(x) = \tan^3(2x)$ at $\frac{\pi}{12}$.
Difficulty level   ● Easy   ○ Normal   ○ Hard
Calculation amount   ○ Small   ● Normal   ○ Large
1) $\frac{4}{3}$
2) $\frac{4}{9}$
3) $\frac{8}{3}$
4) $\frac{8}{9}$

5.7. If the function of $f(x) = |x^3 - 3x + a|$ does not have a derivate at $x = 2$, calculate the value of a.
Difficulty level   ○ Easy   ● Normal   ○ Hard
Calculation amount   ● Small   ○ Normal   ○ Large
1) 2
2) -2
3) 1
4) -1

5.8. Calculate the value of $f'(2) + f'(4)$ if $f(x) = |x^2 - 6|$.
Difficulty level   ○ Easy   ● Normal   ○ Hard
Calculation amount   ● Small   ○ Normal   ○ Large
1) -8
2) 8
3) -4
4) 4

5.9. If $f'(x) = \frac{5}{x}$, calculate the first derivative of $f(x^5)$.
Difficulty level   ○ Easy   ● Normal   ○ Hard
Calculation amount   ● Small   ○ Normal   ○ Large
1) $-\frac{5}{x}$
2) $-\frac{25}{x}$
3) $\frac{25}{x}$
4) $\frac{5}{x^5}$

5.10. If the first derivative of $f(\sin(x))$ is equal to $\cos^3(x)$, determine the value of $f'(x)$.

Difficulty level   ○ Easy   ● Normal   ○ Hard

Calculation amount   ● Small   ○ Normal   ○ Large

1) $1 + x^2$
2) $1 - x^2$
3) x^3
4) $-x^3$

5.11. Calculate the derivative of the function of $f(x) = \text{arc}(\tan(3x))$ at $\frac{1}{3}$.

Difficulty level   ○ Easy   ● Normal   ○ Hard

Calculation amount   ● Small   ○ Normal   ○ Large

1) $\frac{3}{2}$
2) $\frac{4}{3}$
3) $\frac{2}{3}$
4) $\frac{3}{4}$

5.12. If $f\left(\frac{1}{t}\right) + g\left(\sqrt{t}\right) = t^2 + 1$ and $g'(1) = 5$, calculate the value of $f'(1)$.

Difficulty level   ○ Easy   ● Normal   ○ Hard

Calculation amount   ○ Small   ● Normal   ○ Large

1) 1
2) 2
3) $\frac{1}{2}$
4) $-\frac{1}{2}$

5.13. If $2\cos(y) - \sin(x + y) + 2 = 0$, calculate the value of y'_x at $(0, \pi)$.

Difficulty level   ○ Easy   ● Normal   ○ Hard

Calculation amount   ○ Small   ● Normal   ○ Large

1) $\frac{1}{2}$
2) $-\frac{1}{2}$
3) -1
4) 1

5.14. The equation of a curve is given by $x^3 + y^3 = 16$. Calculate the second derivate of y with respect to x.

Difficulty level   ○ Easy   ● Normal   ○ Hard

Calculation amount   ○ Small   ● Normal   ○ Large

1) $-\frac{16y}{x^5}$
2) $\frac{16x}{y^5}$
3) $-\frac{32y}{x^5}$
4) $-\frac{32x}{y^5}$

5.15. If $x = 2 + 3\sin(t)$ and $y = 3 - 2\cos(t)$, calculate the value of y'_x for $t = \frac{\pi}{6}$.

Difficulty level   ○ Easy   ● Normal   ○ Hard

Calculation amount   ○ Small   ● Normal   ○ Large

1) $\frac{2\sqrt{3}}{9}$
2) $\frac{2\sqrt{3}}{3}$
3) $\frac{2\sqrt{2}}{3}$
4) $\frac{4\sqrt{2}}{3}$

5.16. If $x = t^2 + t$ and $y = t^2 - 2t$, calculate the value of $x'_y + y'_x$ for $t = -1$.

1) $\frac{11}{4}$

2) $\frac{13}{4}$

3) $\frac{15}{4}$

4) $\frac{17}{4}$

5.17. Calculate the value of $f'(4)$ if we know that:

$$\lim_{h \to 0} \frac{f(x+h) - f(x-h)}{h} = 2\sqrt{x}$$

1) $\frac{2}{3}$

2) $\frac{4}{3}$

3) 4

4) 2

5.18. Which one of the choices is true about the function of $f(x) = x^2|x|$ at $x = 0$?

1) The first derivative exists, but the second derivative does not.

2) The second derivative exists, but the first derivative does not.

3) The first and second derivatives do not exist.

4) The first and second derivatives exist.

5.19. The function below is differentiable at $x = \frac{\pi}{4}$. Determine the value of b.

$$f(x) = \begin{cases} \sin^2(x) - \cos(2x) & 0 < x \le \frac{\pi}{4} \\ a\tan(x) + b\sin(2x) & \frac{\pi}{4} < x < \frac{\pi}{2} \end{cases}$$

1) -1

2) $-\frac{1}{2}$

3) $\frac{1}{2}$

4) 1

5.20. The function below is differentiable everywhere on $\mathbb{R}$ domain. Determine the value of b.

$$f(x) = \begin{cases} ax + b & x < -1 \\ x^2 + a & x \ge -1 \end{cases}$$

1) 2
2) 1
3) −2
4) −3

5.21. Calculate the derivative of the function below:

$$f(x) = \frac{(2x - 1)^2}{2x^2}$$

Difficulty level   ○ Easy   ● Normal   ○ Hard
Calculation amount   ○ Small   ● Normal   ○ Large
1) $\frac{2x-1}{2x^3}$
2) $\frac{2x-1}{x^3}$
3) $\frac{2x+1}{x^3}$
4) $\frac{2x+1}{2x^3}$

5.22. Calculate the derivative of the function below.

$$f(x) = \frac{\sin(x)}{1 + \tan^2(x)}$$

Difficulty level   ○ Easy   ● Normal   ○ Hard
Calculation amount   ○ Small   ● Normal   ○ Large
1) $\frac{5}{4}$
2) $-\frac{5}{4}$
3) $\frac{5}{8}$
4) $-\frac{5}{8}$

5.23. For the function below, calculate the value of $f'(x = 2)$.

$$f(x) = \left(x^2 - 5x + 6\right)\mathrm{arc}\left(\sin\left(\frac{1}{x}\right)\right)$$

Difficulty level   ○ Easy   ● Normal   ○ Hard
Calculation amount   ○ Small   ● Normal   ○ Large
1) $-\frac{\pi}{3}$
2) $\frac{2\pi}{3}$
3) $-\frac{\pi}{6}$
4) $\frac{\pi}{4}$

5.24. For the following function, calculate the value of $f'(x = -3)$.

$$f(x) = \left(x^2 + 2x - 3\right)\frac{g(x + 2)}{(x^3 + 1)g(2x + 5)}$$

Difficulty level   ○ Easy   ● Normal   ○ Hard
Calculation amount   ○ Small   ● Normal   ○ Large

1) $-\frac{13}{2}$

2) $\frac{13}{2}$

3) $-\frac{2}{13}$

4) $\frac{2}{13}$

5.25. For what value of m the line of $y = 2x + 1$ is tangent to a curve with the following function.

$$y = \frac{-1 + x^2}{m + x}$$

Difficulty level   ○ Easy   ● Normal   ○ Hard
Calculation amount   ○ Small   ● Normal   ○ Large

1) $\frac{3}{4}$

2) $\pm \frac{\sqrt{3}}{8}$

3) $\frac{1}{9}$

4) $\pm \frac{\sqrt{3}}{2}$

5.26. Determine the third derivate of $f(x) = x^4 \, |x|$.

Difficulty level   ○ Easy   ● Normal   ○ Hard
Calculation amount   ○ Small   ● Normal   ○ Large

1) $-60x^2$

2) $60x^2$

3) $-60x|x|$

4) $60x|x|$

5.27. Determine the value of the parameter of "a" if the derivative of $\sqrt{x + a}$ for $x = 2$ is $\frac{1}{4}$.

Difficulty level   ○ Easy   ● Normal   ○ Hard
Calculation amount   ○ Small   ● Normal   ○ Large

1) -2

2) -1

3) 1

4) 2

5.28. Calculate the derivative of $y = \ln e^{\sqrt{\sin (x)}}$ at $x = \frac{\pi}{6}$.

Difficulty level   ○ Easy   ● Normal   ○ Hard
Calculation amount   ○ Small   ● Normal   ○ Large

1) $\frac{\sqrt{3}}{8}$

2) $\frac{\sqrt{6}}{8}$

3) $\frac{\sqrt{3}}{4}$

4) $\frac{\sqrt{6}}{4}$

5.29. Determine the maximum value of the function of $y(x) = x^3 - 3x^2 - 9x + 5$ in the range of $[-2, 2]$.

Difficulty level   ○ Easy   ● Normal   ○ Hard
Calculation amount   ○ Small   ● Normal   ○ Large

1) 9

2) 10

3) 12

4) 17

5.30. Which one of the choices is correct about the function below in its one period.

$$y(x) = \frac{1 - \sin(x)}{\cos(x)}$$

Difficulty level   ○ Easy   ● Normal   ○ Hard
Calculation amount   ○ Small   ● Normal   ○ Large
1) The function is always ascending.
2) The function is always descending.
3) The function has one minimum point.
4) The function has one maximum point.

5.31. Determine the value of $f'(x)g(x) - f(x)g'(x)$ if we have the following functions.

$$f(x) = \left(\sqrt{1+x^2} - x\right)^5, \quad g(x) = \frac{1}{\left(\sqrt{1+x^2} + x\right)^5}$$

Difficulty level   ○ Easy   ● Normal   ○ Hard
Calculation amount   ○ Small   ● Normal   ○ Large
1) −1
2) 0
3) 1
4) 2

5.32. On a curve with the function of $y = x^3 - 6x + 12$, two tangent lines, parallel to x-axis, have been drawn. Determine the distance between these two lines.
Difficulty level   ○ Easy   ○ Normal   ● Hard
Calculation amount   ○ Small   ● Normal   ○ Large
1) 14
2) 6
3) $4\sqrt{2}$
4) $8\sqrt{2}$

5.33. For the function below, calculate the value of $f'(x = -1)$.

$$f(x) = \begin{cases} \dfrac{(x+1)^5}{|x+1|} & x \neq -1 \\ 0 & x = -1 \end{cases}$$

Difficulty level   ○ Easy   ○ Normal   ● Hard
Calculation amount   ○ Small   ● Normal   ○ Large
1) 0
2) 1
3) −1
4) 5

5.34. The point $M(x, y)$ is moving on the curve of $y = \sqrt{x + 8}$. Determine the changing rate of the distance of the point from the origin when $x = 7$.
Difficulty level   ○ Easy   ○ Normal   ● Hard
Calculation amount   ○ Small   ● Normal   ○ Large

1) $\frac{15}{16}$

2) $\frac{15}{8}$

3) $\frac{3}{7}$

4) $\frac{5}{4}$

5.35. Determine the derivative of $f\left(\sqrt{|-x|+3}\right)$ if we have the relation below.

$$\lim_{x \to 2} \frac{f(x) - f(2)}{x - 2} = -\frac{1}{3}$$

Difficulty level   ○ Easy   ○ Normal   ● Hard

Calculation amount   ○ Small   ● Normal   ○ Large

1) $\frac{1}{6}$

2) $\frac{1}{12}$

3) $-\frac{1}{6}$

4) $-\frac{1}{12}$

5.36. Determine the value of $f'(-1)$ if we have:

$$f(x) = \frac{(x+1)h(x)}{(2x+1)h(2x+1)}, \quad h(-1) \neq 0$$

Difficulty level   ○ Easy   ○ Normal   ● Hard

Calculation amount   ○ Small   ● Normal   ○ Large

1) -2

2) -1

3) 1

4) 2

5.37. Determine the value of the parameter of "a" so that the function of $f(x) = \cos^2(x) + \sqrt{3}\sin(x) + a$ has an extremum point with the width of $y = \frac{3}{4}$ in the range of $0 < x < \frac{\pi}{2}$.

Difficulty level   ○ Easy   ○ Normal   ● Hard

Calculation amount   ○ Small   ● Normal   ○ Large

1) 1

2) $\frac{1}{2}$

3) $-\frac{1}{2}$

4) -1

Reference

1. Rahmani-Andebili, M. (2020). Precalculus: Practice problems, methods, and solutions, Springer Nature, 2020.

Solutions of Problems: Derivatives and its Applications

6

Abstract

In this chapter, the problems of the fifth chapter are fully solved, in detail, step-by-step, and with different methods. The subjects include definition of derivative, differentiation formulas, product rule, quotient rule, chain rule, derivatives of trigonometric functions, derivatives of exponential, derivatives of logarithm functions, derivatives of inverse trigonometric functions, derivatives of hyperbolic functions, implicit differentiation, higher-order derivatives, logarithmic differentiation, applications of derivatives, rates of change, critical points, minimum and maximum values, and absolute extrema.

6.1. From the list of derivative rules, we know that [1]:

$$f(x) = e^x \Longrightarrow f'(x) = e^x$$

$$f(x) = u(x)v(x) \Longrightarrow f'(x) = u'(x)v(x) + u(x)v'(x)$$

Based on the information given in the problem, we have:

$$f(x) = xe^x - e^x$$

The problem can be solved as follows.

$$f'(x) = e^x + xe^x - e^x = xe^x$$

$$f'(1) = 1e^1 \Longrightarrow f'(1) = e$$

Choice (4) is the answer.

6.2. From the list of derivative rules, we know that:

$$h(x) = g(u(x)) \Longrightarrow h'(x) = u'(x)g'(u(x))$$

Based on the information given in the problem, we have:

$$f'(1) = 2$$

$$f(x) + g(x^3) = 5x - 1$$

© Springer Nature Switzerland AG 2021
M. Rahmani-Andebili, *Calculus*, https://doi.org/10.1007/978-3-030-64980-7_6

The problem can be solved as follows.

$$f(x) + g(x^3) = 5x - 1 \xRightarrow{\frac{d}{dx}} f'(x) + 3x^2 g'(x^3) = 5 \xRightarrow{x=1} f'(1) + 3g'(1) = 5 \xRightarrow{f'(1) = 2} 2 + 3g'(1) = 5$$

$$\Rightarrow g'(1) = 1$$

Choice (1) is the answer.

6.3. From the list of derivative rules, we know that:

$$f(x) = ax^n \Rightarrow f'(x) = anx^{n-1}$$

A function is ascending in a given range if its derivative is positive. Therefore,

$$y(x) = 1 - 4x^2 \Rightarrow y'(x) = -8x > 0 \Rightarrow x < 0$$

Choice (1) is the answer.

6.4. From the list of derivative rules, we know that:

$$f(x) = \sqrt{x} \Rightarrow f'(x) = \frac{1}{2\sqrt{x}}$$

First, we should simplify the function as follows:

$$f(x) = \frac{x - \sqrt{x}}{1 - \sqrt{x}} = \frac{\sqrt{x}(\sqrt{x} - 1)}{1 - \sqrt{x}} = -\sqrt{x}$$

Therefore,

$$\Rightarrow f'(x) = -\frac{1}{2\sqrt{x}} \Rightarrow f'\left(\frac{1}{4}\right) = -\frac{1}{2\sqrt{\frac{1}{4}}} = -1$$

Choice (1) is the answer.

6.5. From the list of derivative rules, we know that:

$$f(x) = u^n(x) \Rightarrow f'(x) = nu'(x)u^{n-1}(x)$$

Therefore

$$f(x) = \left(x^{100} + x^{50} + 50x^2 + 50x + 1\right)^{10}$$

$$\Rightarrow f'(x) = 10\left(100x^{99} + 50x^{49} + 100x + 50\right)\left(x^{100} + x^{50} + 50x^2 + 50x + 1\right)^9$$

$$\Rightarrow f'(0) = 10(0 + 0 + 0 + 50)(0 + 0 + 0 + 0 + 1)^9 = 500$$

Choice (4) is the answer.

6.6. From the list of derivative rules, we know that:

$$f(x) = \tan^n(u(x)) \Rightarrow f'(x) = nu'(x)\big(1 + \tan^2(u(x))\big)\tan^{n-1}(u(x))$$

Therefore,

$$f(x) = \tan^3(2x) \Rightarrow f'(x) = 3 \times 2\big(1 + \tan^2(2x)\big)\tan^2(2x)$$

$$\Rightarrow f'\left(\frac{\pi}{12}\right) = 3 \times 2\left(1 + \tan^2\left(\frac{\pi}{6}\right)\right)\tan^2\left(\frac{\pi}{6}\right) = 6 \times \left(1 + \frac{1}{3}\right) \times \frac{1}{3} = \frac{8}{3}$$

Choice (3) is the answer.

6.7. Based on the information given in the problem, we have:

$$f(x) = \left|x^3 - 3x + a\right|$$

The derivative of an absolute value equation does not exist at its roots. Therefore, we need to solve the equation below:

$$f(2) = 0$$

$$\Rightarrow \left|2^3 - 3 \times 2 + a\right| = 0 \Rightarrow 8 - 6 + a = 0 \Rightarrow 2 + a = 0 \Rightarrow a = -2$$

Choice (2) is the answer.

6.8. From the list of derivative rules, we know that:

$$f(x) = |u(x)| \Rightarrow f'(x) = \frac{u'(x)u(x)}{|u(x)|}$$

Based on the information given in the problem, we have:

$$f(x) = \left|x^2 - 6\right|$$

Therefore,

$$\Rightarrow f'(x) = \frac{2x \times (x^2 - 6)}{|x^2 - 6|} \Rightarrow f'(2) = \frac{4 \times (4 - 6)}{|4 - 6|} = -4, \quad f'(4) = \frac{8 \times (16 - 6)}{|16 - 6|} = 8$$

$$\Rightarrow f'(2) + f'(4) = -4 + 8 = 4$$

Choice (4) is the answer.

6.9. From the list of derivative rules, we know that:

$$f(x) = g(h(x)) \Rightarrow f'(x) = h'(x)g'(h(x))$$

Based on the information given in the problem, we have:

$$f'(x) = \frac{5}{x}$$

The problem can be solved as follows.

$$(f(x^5))' = 5x^4 \times f'(x^5) = 5x^4 \times \frac{5}{x^5} \Rightarrow (f(x^5))' = \frac{25}{x}$$

Choice (3) is the answer.

6.10. From the list of derivative rules, we have:

$$(f(g(x)))' = g'(x)f'(g(x))$$

Based on the information given in the problem, we have:

$$(f(\sin(x)))' = \cos^3(x)$$

The problem can be solved as follows.

$$(f(\sin(x)))' = \cos^3(x) \Rightarrow \cos(x) \times f'(\sin(x)) = \cos^3(x) \Rightarrow f'(\sin(x)) = \cos^2(x)$$

$$\Rightarrow f'(\sin(x)) = 1 - \sin^2(x) \Rightarrow f'(x) = 1 - x^2$$

Choice (2) is the answer.

6.11. From the list of derivative rules, we know that:

$$f(x) = \text{arc}(\tan(u(x))) \Rightarrow f'(x) = \frac{u'(x)}{1 + u^2(x)}$$

Therefore,

$$f(x) = \text{arc}(\tan(3x)) \Rightarrow f'(x) = \frac{3}{1 + 9x^2}$$

$$f'\left(\frac{1}{3}\right) = \frac{3}{1 + 9\left(\frac{1}{3}\right)^2} = \frac{3}{2}$$

Choice (1) is the answer.

6.12. From the list of derivative rules, we know that:

$$f(x) = ax^n \Rightarrow f'(x) = anx^{n-1}$$

$$f(x) = g(h(x)) \Rightarrow f'(x) = h'(x)g'(h(x))$$

Based on the information given in the problem, we have:

$$f\left(\frac{1}{t}\right) + g(\sqrt{t}) = t^2 + 1$$

$$g'(1) = 5$$

The problem can be solved as follows.

$$f\left(\frac{1}{t}\right) + g\left(\sqrt{t}\right) = t^2 + 1 \xrightarrow{\frac{d}{dx}} \left(-\frac{1}{t^2}\right) \times f'\left(\frac{1}{t}\right) + \frac{1}{2\sqrt{t}} \times g'\left(\sqrt{t}\right) = 2t$$

$$t = 1 \Rightarrow -f'(1) + \frac{1}{2}g'(1) = 2 \xrightarrow{g'(1) = 5} -f'(1) + \frac{5}{2} = 2 \Rightarrow f'(1) = \frac{1}{2}$$

Choice (3) is the answer.

6.13. From the derivative rules, we know that:

$$f(x,y) = 0 \Rightarrow y'_x = -\frac{f'_x(x,y)}{f'_y(x,y)} = -\frac{\frac{d}{dx}f(x,y)}{\frac{d}{dy}f(x,y)}$$

Based on the information given in the problem, we have:

$$2\cos(y) - \sin(x+y) + 2 = 0$$

The problem can be solved as follows.

$$y'_x = -\frac{\frac{d}{dx}(2\cos(y) - \sin(x+y) + 2)}{\frac{d}{dy}(2\cos(y) - \sin(x+y) + 2)} = -\frac{-\cos(x+y)}{-2\sin(y) - \cos(x+y)} = -\frac{\cos(x+y)}{2\sin(y) + \cos(x+y)}$$

$$\xrightarrow{(x,y) = (0,\pi)} y'_x = -\frac{-1}{0-1} \Rightarrow y'_x = -1$$

Choice (3) is the answer.

6.14. From the derivative rules, we know that:

$$f(x,y) = 0 \Rightarrow y'_x = -\frac{f'_x(x,y)}{f'_y(x,y)} = -\frac{\frac{d}{dx}f(x,y)}{\frac{d}{dy}f(x,y)}$$

Based on the information given in the problem, we have:

$$x^3 + y^3 = 16$$

The problem can be solved as follows.

$$y'_x = y' = -\frac{3x^2}{3y^2} = -\frac{x^2}{y^2}$$

$$\xrightarrow{y' = -\frac{x^2}{y^2}} y'' = -\frac{2xy^2 - 2yy'x^2}{y^4} \xrightarrow{\quad} y'' = -\frac{2xy^2 - 2y\left(-\frac{x^2}{y^2}\right)x^2}{y^4} = -\frac{2xy^3 + 2x^4}{y^5} = -\frac{2x(y^3 + x^3)}{y^5}$$

$$\xrightarrow{x^3 + y^3 = 16} y'' = -\frac{2x \times 16}{y^5} \Rightarrow y'' = -\frac{32x}{y^5}$$

Choice (4) is the answer.

6.15. From the derivative rules, we know that:

$$\begin{cases} y = y(t) \\ x = x(t) \end{cases} \Rightarrow y'_x = \frac{y'_t}{x'_t} = \frac{\frac{d}{dt}y(t)}{\frac{d}{dt}x(t)}$$

Based on the information given in the problem, we have:

$$x = 2 + 3\sin(t)$$

$$y = 3 - 2\cos(t)$$

Hence,

$$y'_x = \frac{y'_t}{x'_t} = \frac{2\sin(t)}{3\cos(t)} = \frac{2}{3}\tan(t)$$

$$t = \frac{\pi}{6} \Rightarrow y'_x = \frac{2}{3} \times \frac{\sqrt{3}}{3} \Rightarrow y'_x = \frac{2\sqrt{3}}{9}$$

Choice (1) is the answer.

6.16. From the derivative rules, we know that:

$$\begin{cases} y = y(t) \\ x = x(t) \end{cases} \Rightarrow y'_x = \frac{y'_t}{x'_t} = \frac{\frac{d}{dt}y(t)}{\frac{d}{dt}x(t)}, x'_y = \frac{x'_t}{y'_t} = \frac{\frac{d}{dt}x(t)}{\frac{d}{dt}y(t)}$$

Based on the information given in the problem, we have:

$$x = t^2 + t$$

$$y = t^2 - 2t$$

Therefore,

$$\Rightarrow \begin{cases} y'_x = \frac{y'_t}{x'_t} = \frac{2t-2}{2t+1} \\ x'_y = \frac{x'_t}{y'_t} = \frac{2t+1}{2t-2} \end{cases} \xrightarrow{t=-1} \begin{cases} y'_x = \frac{-2-2}{-2+1} = 4 \\ x'_y = \frac{-2+1}{-2-2} = \frac{1}{4} \end{cases} \Rightarrow y'_x + x'_x = 4 + \frac{1}{4} \Rightarrow y'_x + x'_x = \frac{17}{4}$$

Choice (4) is the answer.

6.17. Based on the information given in the problem, we have:

$$\lim_{h \to 0} \frac{f(x+h) - f(x-h)}{h} = 2\sqrt{x} \qquad (1)$$

From definition of derivative, we know:

$$f'(x) = \lim_{h \to 0} \frac{f(x+h) - f(x)}{h} = \lim_{h \to 0} \frac{f(x) - f(x-h)}{h}$$

The problem can be solved as follows.

$$\lim_{h \to 0} \frac{f(x+h) - f(x-h)}{h} = \lim_{h \to 0} \frac{f(x+h) - f(x) + f(x) - f(x-h)}{h}$$

$$= \lim_{h \to 0} \frac{f(x+h) - f(x)}{h} + \lim_{h \to 0} \frac{f(x) - f(x-h)}{h} = 2f'(x) \tag{2}$$

Solving (1) and (2):

$$2f'(x) = 2\sqrt{x} \Rightarrow f'(x) = \sqrt{x} \Rightarrow f'(4) = \sqrt{4} = 2$$

Choice (4) is the answer.

6.18. The function can be simplified as follows:

$$f(x) = x^2 |x| = \begin{cases} x^3 & x \geq 0 \\ -x^3 & x < 0 \end{cases}$$

$$\Rightarrow f(0) = \begin{cases} 0 & x \geq 0 \\ 0 & x < 0 \end{cases} \Rightarrow f(0^-) = f(0^+) \tag{1}$$

Now, we can determine its first and second derivatives as follows:

$$\Rightarrow f'(x) = \begin{cases} 3x^2 & x \geq 0 \\ -3x^2 & x < 0 \end{cases} \Rightarrow f'(0) = \begin{cases} 0 & x \geq 0 \\ 0 & x < 0 \end{cases} \Rightarrow f'(0^-) = f'(0^+) \tag{2}$$

$$\Rightarrow f''(x) = \begin{cases} 6x & x \geq 0 \\ -6x & x < 0 \end{cases} \Rightarrow f''(0) = \begin{cases} 0 & x \geq 0 \\ 0 & x < 0 \end{cases} \Rightarrow f''(0^-) = f''(0^+) \tag{3}$$

From (1) and (2), we can conclude that the first derivative of the function exits. Likewise, from (1), (2), and (3), we can say that the second derivative of the function does exist. Choice (4) is the answer.

6.19. Based on the information given in the problem, we have:

$$f(x) = \begin{cases} \sin^2(x) - \cos(2x) & 0 < x \leq \dfrac{\pi}{4} \\ a\tan(x) + b\sin(2x) & \dfrac{\pi}{4} < x < \dfrac{\pi}{2} \end{cases} \tag{1}$$

The function is differentiable at $x = \frac{\pi}{4}$. Therefore, we can conclude that:

$$f\left(\left(\frac{\pi}{4}\right)^-\right) = f\left(\left(\frac{\pi}{4}\right)^+\right) \tag{2}$$

$$f'\left(\left(\frac{\pi}{4}\right)^-\right) = f'\left(\left(\frac{\pi}{4}\right)^+\right) \tag{3}$$

Solving (1) and (2):

$$\sin^2\left(\frac{\pi}{4}\right) - \cos\left(2 \times \frac{\pi}{4}\right) = a\tan\left(\frac{\pi}{4}\right) + b\sin\left(2 \times \frac{\pi}{4}\right) \Rightarrow a + b = \frac{1}{2} \tag{4}$$

Solving (1) and (3):

$$2\sin\left(\frac{\pi}{4}\right)\cos\left(\frac{\pi}{4}\right) + 2\sin\left(2\times\frac{\pi}{4}\right) = a\left(1 + \tan^2\left(\frac{\pi}{4}\right)\right) + 2b\cos\left(2\times\frac{\pi}{4}\right)$$

$$\Rightarrow 1 + 2 = 2a + 0 \Rightarrow a = \frac{3}{2} \tag{5}$$

Solving (4) and (5):

$$b = -1$$

Choice (1) is the answer.

6.20. Based on the information given in the problem, we have:

$$f(x) = \begin{cases} ax + b & x < -1 \\ x^2 + a & x \geq -1 \end{cases} \tag{1}$$

The function is differentiable everywhere on $\mathbb{R}$ domain including $x = -1$. Hence:

$$f\big((-1)^-\big) = f\big((-1)^+\big) \tag{2}$$

$$f'\big((-1)^-\big) = f'\big((-1)^+\big) \tag{3}$$

Solving (1) and (2):

$$-a + b = 1 + a \Rightarrow -2a + b = 1 \tag{4}$$

Solving (1) and (3):

$$a = -2 \tag{5}$$

Solving (4) and (5):

$$b = -3$$

Choice (4) is the answer.

6.21. From list of derivative rules, we know that:

$$f(x) = \frac{g(x)}{h(x)} \Rightarrow f'(x) = \frac{g'(x)h(x) - h'(x)g(x)}{h^2(x)}$$

Therefore,

$$f(x) = \frac{(2x-1)^2}{2x^2} \Rightarrow f'(x) = \frac{4(2x-1)(2x^2) - 4x(2x-1)^2}{4x^4}$$

$$\Rightarrow f'(x) = \frac{16x^3 - 8x^2 - 16x^3 + 16x^2 - 4x}{4x^4} = \frac{8x^2 - 4x}{4x^4} = \frac{2x-1}{x^3}$$

Choice (2) is the answer.

6.22. From the list of derivative rules, we know that:

$$f(x) = u(x)v(x) \Rightarrow f'(x) = u'(x)v(x) + u(x)v'(x)$$

From trigonometry, we have:

$$1 + \tan^2(x) = \frac{1}{\cos^2(x)}$$

Based on the information given in the problem, we have:

$$f(x) = \frac{\sin(x)}{1 + \tan^2(x)} \Rightarrow f(x) = \frac{\sin(x)}{\frac{1}{\cos^2(x)}} = \sin(x)\cos^2(x)$$

$$\Rightarrow f'(x) = \cos(x)\cos^2(x) - \sin(x) \times 2\sin(x)\cos(x) = \cos^3(x) - 2\sin^2(x)\cos(x)$$

$$\Rightarrow f'\left(\frac{\pi}{3}\right) = \left(\frac{1}{2}\right)^3 - 2\left(\frac{\sqrt{3}}{2}\right)^2 \frac{1}{2} = \frac{1}{8} - \frac{3}{4} = -\frac{5}{8}$$

Choice (4) is the answer.

6.23. From the list of derivative rules, we know that:

$$f(x) = u(x)v(x) \Rightarrow f'(x) = u'(x)v(x) + u(x)v'(x)$$

$$f(x) = ax^n \Rightarrow f'(x) = anx^{n-1}$$

From trigonometry, we know that:

$$\text{arc}\left(\sin\left(\frac{1}{2}\right)\right) = \frac{\pi}{6}$$

Based on the information given in the problem, we have:

$$f(x) = \left(x^2 - 5x + 6\right)\text{arc}\left(\sin\left(\frac{1}{x}\right)\right)$$

Therefore,

$$f'(x) = (2x - 5)\text{arc}\left(\sin\left(\frac{1}{x}\right)\right) + \left(x^2 - 5x + 6\right)\left(\text{arc}\left(\sin\left(\frac{1}{x}\right)\right)\right)'$$

$$f'(2) - (2 \times 2 - 5)\text{arc}\left(\sin\left(\frac{1}{2}\right)\right) + 0 \times \left(\text{arc}\left(\sin\left(\frac{1}{x}\right)\right)\right)' - (-1) \times \frac{\pi}{6} \rightarrow f'(2) - -\frac{\pi}{6}$$

As can be seen, we did not need to calculate the value of $\left(\text{arc}\left(\sin\left(\frac{1}{x}\right)\right)\right)'$. Choice (3) is the answer.

6.24. From the list of derivative rules, we know that:

$$f(x) = u(x)v(x) \Rightarrow f'(x) = u'(x)v(x) + u(x)v'(x)$$

$$f(x) = ax^n \Rightarrow f'(x) = anx^{n-1}$$

Based on the information given in the problem, we have:

$$f(x) = \left(x^2 + 2x - 3\right) \times \frac{g(x+2)}{(x^3 + 1)g(2x + 5)} ⏟u(x) \times v(x)$$

The problem can be solved as follows.

$$f'(x) = (2x + 2)\frac{g(x+2)}{(x^3 + 1)g(2x + 5)} + \left(x^2 + 2x - 3\right)v'(x)$$

$$\Rightarrow f'(x = -3) = (-4) \times \frac{g(-1)}{-26g(-1)} + (0) \times v'(x = -3) = -4 \times \left(-\frac{1}{26}\right)$$

$$\Rightarrow f'(x = -3) = \frac{2}{13}$$

As can be seen, we did not need to calculate the value of $v'(x = -3)$. Choice (4) is the answer.

6.25. Since the line is tangent to the curve, equating their equations and solving them will result in a new equation that will have repeated roots. In other words, the discriminant of the new equation must be zero ($\Delta = 0$).

$$\frac{-1 + x^2}{m + x} = 2x + 1 \Rightarrow -1 + x^2 = 2x^2 + x + 2mx + m \Rightarrow x^2 + (2m + 1)x + m + 1 = 0$$

$$\Delta = 0 \Rightarrow (2m + 1)^2 - 4(m + 1) = 0 \Rightarrow 4m^2 - 3 = 0 \Rightarrow m = \pm\frac{\sqrt{3}}{2}$$

Choice (4) is the answer.

6.26. From the list of derivative rules, we know that:

$$f(x) = ax^n \Rightarrow f'(x) = anx^{n-1}$$

Based on the information given in the problem, we have:

$$f(x) = x^4|x|$$

Therefore,

$$f(x) = \begin{cases} x^5 & ,x \geq 0 \\ -x^5 & ,x < 0 \end{cases} \Rightarrow f'(x) = \begin{cases} 5x^4 & ,x > 0 \\ 0 & ,x = 0 \\ -5x^4 & ,x < 0 \end{cases} \Rightarrow f''(x) = \begin{cases} 20x^3 & ,x > 0 \\ 0 & ,x = 0 \\ -20x^3 & ,x < 0 \end{cases}$$

$$\Rightarrow f'''(x) = \begin{cases} 60x^2 & ,x > 0 \\ 0 & ,x = 0 \\ -60x^2 & ,x < 0 \end{cases} \Rightarrow f''''(x) = 60x|x|$$

Choice (4) is the answer.

6.27. Based on the information given in the problem, we have:

$$f(x) = \sqrt{x+a} \Longrightarrow f'(2) = \frac{1}{4} \tag{1}$$

From the list of derivative rules, we know that:

$$f(x) = \sqrt{u(x)} \Longrightarrow f'(x) = \frac{u'(x)}{2\sqrt{u(x)}} \tag{2}$$

Solving (1) and (2):

$$\left.\frac{1}{2\sqrt{x+a}}\right|_{x=2} = \frac{1}{4} \Longrightarrow \frac{1}{2\sqrt{2+a}} = \frac{1}{4} \Longrightarrow \sqrt{2+a} = 2 \Longrightarrow a = 2$$

Choice (4) is the answer.

6.28. First, we should simplify the function as follows:

$$y(x) = \ln e^{\sqrt{\sin(x)}} = \sqrt{\sin(x)}\ln e = \sqrt{\sin(x)} \tag{1}$$

From the list of derivative rules, we know that:

$$f(x) = \sqrt{u(x)} \Longrightarrow f'(x) = \frac{u'(x)}{2\sqrt{u(x)}} \tag{2}$$

Solving (1) and (2):

$$\Longrightarrow y'(x) = \frac{\cos(x)}{2\sqrt{\sin(x)}} \Longrightarrow y'\left(\frac{\pi}{6}\right) = \frac{\cos\left(\frac{\pi}{6}\right)}{2\sqrt{\sin\left(\frac{\pi}{6}\right)}} = \frac{\frac{\sqrt{3}}{2}}{2\sqrt{\frac{1}{2}}} = \frac{\sqrt{6}}{4}$$

Choice (4) is the answer.

6.29. From the list of derivative rules, we know that:

$$f(x) = ax^n \Longrightarrow f'(x) = anx^{n-1}$$

To determine the maximum value of a function for a given range, we need to calculate the value of the function at its critical points including the extremum points and the beginning and the end of the range.

$$y(x) = x^3 - 3x^2 - 9x + 5 \Longrightarrow y'(x) = 3x^2 - 6x - 9$$

To find the extremum points of the function:

$$\Longrightarrow y'(x) = 0 \Longrightarrow 3x^2 - 6x - 9 = 0 \Longrightarrow x^2 - 2x - 3 = 0 \Longrightarrow x = 3, -1$$

$x = 3$ is not acceptable because it is out of the range. The value of the function at the critical points can be calculated as follows:

$$y(-2) = (-2)^3 - 3(-2)^2 - 9(-2) + 5 = 3$$

$$y(-1) = (-1)^3 - 3(-1)^2 - 9(-1) + 5 = 10$$

$$y(2) = (2)^3 - 3(2)^2 - 9(2) + 5 = -17$$

Therefore, the maximum value of the function is 10. Choice (2) is the answer.

6.30. From the list of derivative rules, we know that:

$$f(x) = \frac{g(x)}{h(x)} \Rightarrow f'(x) = \frac{g'(x)h(x) - h'(x)g(x)}{h^2(x)} \tag{1}$$

First, we need to take its derivative as follows:

$$y(x) = \frac{1 - \sin(x)}{\cos(x)} \Rightarrow y'(x) = \frac{-\cos(x)\cos(x) - (-\sin(x))(1 - \sin(x))}{\cos^2(x)}$$

$$y'(x) = \frac{-\cos^2(x) + \sin(x) - \sin^2(x)}{\cos^2(x)} = \frac{-1 + \sin(x)}{\cos^2(x)}$$

The $y'(x)$ is always nonpositive because $\sin(x) \leq 1$. Hence, the function is a descending function. Choice (2) is the answer.

6.31. Based on the information given in the problem, we have:

$$f(x) = \left(\sqrt{1 + x^2} - x\right)^5, \quad g(x) = \frac{1}{\left(\sqrt{1 + x^2} + x\right)^5} \tag{1}$$

From the list of derivative rules, we know that:

$$\left(\frac{f(x)}{g(x)}\right)' = \frac{f'(x)g(x) - g'(x)f(x)}{(g(x))^2} \Rightarrow f'(x)g(x) - g'(x)f(x) = \left(\frac{f(x)}{g(x)}\right)'(g(x))^2 \tag{2}$$

Solving (1) and (2):

$$f'(x)g(x) - g'(x)f(x) = \left(\frac{\left(\sqrt{1 + x^2} - x\right)^5}{\frac{1}{\left(\sqrt{1+x^2}+x\right)^5}}\right)'\left(\frac{1}{\left(\sqrt{1 + x^2} + x\right)^5}\right)^2$$

$$= \left((1 + x^2 - x^2)^5\right)'\frac{1}{\left(\sqrt{1 + x^2} + x\right)^{10}} = (1)' \times \frac{1}{\left(\sqrt{1 + x^2} + x\right)^{10}} = 0 \times \frac{1}{\left(\sqrt{1 + x^2} + x\right)^{10}} = 0$$

Choice (2) is the answer.

6.32. Since the tangent lines are parallel to x-axis, their slope angles must be zero. Therefore,

$$y = x^3 - 6x + 12 \Rightarrow y' = 3x^2 - 6 = 0 \Rightarrow x = \sqrt{2}, -\sqrt{2}$$

$$x_1 = \sqrt{2} \Rightarrow y_1 = \left(\sqrt{2}\right)^3 - 6\sqrt{2} + 12 = 2\sqrt{2} - 6\sqrt{2} + 12 = -4\sqrt{2} + 12$$

$$x_2 = -\sqrt{2} \Rightarrow y_2 = \left(-\sqrt{2}\right)^3 - 6\left(-\sqrt{2}\right) + 12 = -2\sqrt{2} + 6\sqrt{2} + 12 = 4\sqrt{2} + 12$$

$$y_2 - y_1 = \left(4\sqrt{2} + 12\right) - \left(-4\sqrt{2} + 12\right) \Rightarrow y_2 - y_1 = 8\sqrt{2}$$

Choice (4) is the answer.

6.33. Based on the information given in the problem, we have:

$$f(x) = \begin{cases} \dfrac{(x+1)^5}{|x+1|} & x \neq -1 \\ 0 & x = -1 \end{cases}$$

This problem can be solved by using the definition of derivative of a function as follows.

$$f'(x_0) = \lim_{x \to x_0} \frac{f(x) - f(x_0)}{x - x_0}$$

$$\Rightarrow f'(-1) = \lim_{x \to (-1)} \frac{\frac{(x+1)^5}{|x+1|} - 0}{x - (-1)} = \lim_{x \to (-1)} \frac{\frac{(x+1)^5}{|x+1|}}{x + 1} = \lim_{x \to (-1)} \frac{(x+1)^4}{|x+1|} = \lim_{x \to (-1)} |x+1|^3 = 0$$

Choice (1) is the answer.

6.34. From the list of derivative rules, we know that:

$$f(x) = \sqrt{u(x)} \Rightarrow f'(x) = \frac{u'(x)}{2\sqrt{u(x)}} \tag{1}$$

The distance of the point from the origin can be calculated as follows:

$$D(x) = \sqrt{(x-0)^2 + \left(\sqrt{x+8} - 0\right)^2} = \sqrt{x^2 + x + 8}$$

In addition, the changing rate of the distance can be determined as follows:

$$D'(x) = \frac{d}{dx} D(x) = \frac{d}{dx} \sqrt{x^2 + x + 8} \tag{2}$$

Solving (1) and (2):

$$D'(x) = \frac{2x+1}{2\sqrt{x^2 + x + 8}} \Rightarrow D'(7) = \frac{2 \times 7 + 1}{2\sqrt{7^2 + 7 + 8}} = \frac{15}{16}$$

Choice (1) is the answer.

6.35. From the list of derivative rules, we know that:

$$f(x) = g(h(x)) \Rightarrow f'(x) = h'(x)g'(h(x))$$

$$f(x) = \sqrt{u(x)} \Rightarrow f'(x) = \frac{u'(x)}{2\sqrt{u(x)}}$$

The problem should be solved by using the definition of derivative of a function as follows.

$$f'(x_0) = \lim_{x \to x_0} \frac{f(x) - f(x_0)}{x - x_0} \tag{1}$$

Based on the information given in the problem, we have:

$$\lim_{x \to 2} \frac{f(x) - f(2)}{x - 2} = -\frac{1}{3} \tag{2}$$

Solving (1) and (3):

$$f'(2) = -\frac{1}{3} \tag{3}$$

Therefore,

$$\frac{d}{dx}\left(f\left(\sqrt{|-x|+3}\right)\right)\Big|_{x=-1} = \frac{d}{dx}\left(f(\sqrt{-x+3})\right)\Big|_{x=-1} \Rightarrow \frac{-1}{2\sqrt{-x+3}} f'(\sqrt{-x+3})\Big|_{x=-1} = -\frac{1}{4} f'(2) \tag{4}$$

Solving (3) and (4):

$$\frac{d}{dx}\left(f\left(\sqrt{|-x|+3}\right)\right)\Big|_{x=-1} = -\frac{1}{4} \times \left(-\frac{1}{3}\right) = \frac{1}{12}$$

Choice (2) is the answer.

6.36. Based on the information given in the problem, we have:

$$f(x) = \frac{(x+1)h(x)}{(2x+1)h(2x+1)}, \quad h(-1) \neq 0 \tag{1}$$

The derivative of this function should be solved by using the definition of derivative of a function as follows:

$$f'(x_0) = \lim_{x \to x_0} \frac{f(x) - f(x_0)}{x - x_0} \Rightarrow f'(-1) = \lim_{x \to -1} \frac{f(x) - f(-1)}{x - (-1)} \tag{2}$$

Solving (1) and (2):

$$f'(-1) = \lim_{x \to -1} \frac{\frac{(x+1)h(x)}{(2x+1)h(2x+1)} - \frac{(-1+1)h(-1)}{(-2+1)h(-2+1)}}{x - (-1)} = \lim_{x \to -1} \frac{\frac{(x+1)h(x)}{(2x+1)h(2x+1)} - 0}{x + 1}$$

$$f'(-1) = \lim_{x \to -1} \frac{h(x)}{(2x+1)h(2x+1)} = \frac{h(-1)}{(-1)h(-1)} = -1$$

Choice (2) is the answer.

6.37. Based on the information given in the problem, the width of the extremum point is as follows:

$$y(x_M) = \frac{3}{4} \tag{1}$$

To determine the extremum points of a function, we need to find the roots of the derivative of the function as follows.

$$f'(x) = 0 \tag{2}$$

$$f(x) = \cos^2(x) + \sqrt{3}\sin(x) + a \tag{3}$$

$$\Rightarrow f'(x) = -2\sin(x)\cos(x) + \sqrt{3}\cos(x) = \cos(x)\left(-2\sin(x) + \sqrt{3}\right) \tag{4}$$

Solving (2) and (4):

$$\cos(x)\left(-2\sin(x) + \sqrt{3}\right) = 0 \Rightarrow \begin{cases} \cos(x) = 0 & (5) \\ \sin(x) = \dfrac{\sqrt{3}}{2} & (6) \end{cases}$$

There is no answer for equation (5) in the range of $0 < x < \frac{\pi}{2}$. However, $x = \frac{\pi}{3}$ is only answer for equation (6).

Therefore, by using $x_M = \frac{\pi}{3}$ and (1) in (3), we have:

$$\frac{3}{4} = \cos^2\left(\frac{\pi}{3}\right) + \sqrt{3}\sin\left(\frac{\pi}{3}\right) + a \Rightarrow \frac{3}{4} = \frac{1}{4} + \frac{3}{2} + a \Rightarrow a = -1$$

Choice (4) is the answer.

Reference

1. Rahmani-Andebili, M. (2020). Precalculus: Practice problems, methods, and solutions, Springer Nature, 2020.

Problems: Definite and Indefinite Integrals

Abstract

In this chapter, the basic and advanced problems of definite and indefinite integrals are presented. The subjects include definite integrals, indefinite integrals, substitution rule for integrals, integration techniques, integration by parts, integrals involving trigonometric functions, trigonometric substitutions, integration using partial fractions, integrals involving roots, integrals involving quadratics, applications of integrals, average value, area between curves, and volume of solid of revolution. To help students study the chapter in the most efficient way, the problems are categorized based on their difficulty levels (easy, normal, and hard) and calculation amounts (small, normal, and large). Moreover, the problems are ordered from the easiest problem with the smallest computations to the most difficult problems with the largest calculations.

7.1. Calculate the value of the definite integral below [1].

$$\int_1^2 \frac{x+4}{x^3}\,dx$$

Difficulty level	● Easy	○ Normal	○ Hard
Calculation amount	● Small	○ Normal	○ Large

1) 1
2) 2
3) 3
4) 4

7.2. Solve the following indefinite integral.

$$\int \left(e^x + 2xe^{x^2} \right) dx$$

Difficulty level	● Easy	○ Normal	○ Hard
Calculation amount	● Small	○ Normal	○ Large

1) $-e^x - e^{x^2} + c$
2) $-e^x + e^{x^2} + c$
3) $e^x - e^{x^2} + c$
4) $e^x + e^{x^2} + c$

M. Rahmani-Andebili, *Calculus*, https://doi.org/10.1007/978-3-030-64980-7_7

7.3. Calculate the value of $f''(1)$ if we know that:

$$f(x) = \int (x^3 + 5x)\, dx$$

Difficulty level   ● Easy   ○ Normal   ○ Hard
Calculation amount   ● Small   ○ Normal   ○ Large
1) 2
2) 4
3) 6
4) 8

7.4. Calculate the value of $f''\left(\frac{\pi}{2}\right)$ if $f(x) = \int \cos^3(x)\, dx$.

Difficulty level   ● Easy   ○ Normal   ○ Hard
Calculation amount   ● Small   ○ Normal   ○ Large
1) 0
2) 1
3) −1
4) $\frac{3\sqrt{2}}{2}$

7.5. Calculate the value of the definite integral below.

$$\int_{-2}^{-1} \frac{x^3 + x^2 - 1}{x^2}\, dx$$

Difficulty level   ● Easy   ○ Normal   ○ Hard
Calculation amount   ○ Small   ● Normal   ○ Large
1) 1
2) −1
3) $\frac{1}{2}$
4) $-\frac{1}{2}$

7.6. Solve the indefinite integral below.

$$\int \frac{x - 2}{\sqrt{x}}\, dx$$

Difficulty level   ● Easy   ○ Normal   ○ Hard
Calculation amount   ○ Small   ● Normal   ○ Large
1) $\frac{2}{3}\sqrt{x}(x + 6) + c$
2) $\frac{2}{3}\sqrt{x}(x - 6) + c$
3) $\frac{1}{3}\sqrt{x}(x + 6) + c$
4) $\frac{2}{3}\sqrt{x}(x - 6) + c$

7.7. Calculate the value of the following definite integral.

$$\int_0^1 \frac{x^2}{(x^3 + 1)^4} dx$$

Difficulty level   ● Easy   ○ Normal   ○ Hard
Calculation amount   ○ Small   ● Normal   ○ Large
1) $\frac{5}{36}$
2) $\frac{7}{36}$
3) $\frac{5}{72}$
4) $\frac{7}{72}$

7.8. Calculate the integral of the function below for the range of $\infty < x < +\infty$.

$$f(x) = \frac{1}{x^2 + 4}$$

Difficulty level   ○ Easy   ● Normal   ○ Hard
Calculation amount   ● Small   ○ Normal   ○ Large
1) $\frac{\pi}{2}$
2) π
3) $\frac{3\pi}{2}$
4) 2π

7.9. Calculate the value of the definite integral below.

$$\int_1^4 \frac{x - \sqrt{x}}{\sqrt{x}} dx$$

Difficulty level   ○ Easy   ● Normal   ○ Hard
Calculation amount   ● Small   ○ Normal   ○ Large
1) $\frac{1}{3}$
2) $\frac{2}{3}$
3) $\frac{4}{3}$
4) $\frac{5}{3}$

7.10. If the primary function of $f(x)$ is equal to $\frac{x^3}{6}$, determine the first derivate of $f(\frac{1}{x})$ with respect to x.

Difficulty level   ○ Easy   ● Normal   ○ Hard
Calculation amount   ● Small   ○ Normal   ○ Large
1) $-\frac{1}{x^3}$
2) $-\frac{x}{6}$
3) $-\frac{1}{2}$
4) $\frac{1}{x^3}$

7.11. Calculate the value of $F'(\lambda = 0)$ if:

$$F(\lambda) = \int_0^{\lambda} \frac{1}{x^4 + 2} dx$$

Difficulty level   ○ Easy   ● Normal   ○ Hard
Calculation amount   ● Small   ○ Normal   ○ Large
1) 1
2) $\frac{1}{2}$
3) $\frac{1}{3}$
4) 0

7.12. Calculate the value of the definite integral below.

$$\int_{-1}^{1} \frac{x^2}{1 + x^2} \text{arc}(\tan(x)) dx$$

Difficulty level   ○ Easy   ● Normal   ○ Hard
Calculation amount   ● Small   ○ Normal   ○ Large
1) 0
2) 1
3) $\frac{\pi}{4}$
4) $\frac{\pi}{2}$

7.13. Calculate the integral of the function below for the range of $-\frac{1}{2} < x < \frac{1}{2}$.

$$f(x) = \frac{1}{\sqrt{1 - x^2}}$$

Difficulty level   ○ Easy   ● Normal   ○ Hard
Calculation amount   ● Small   ○ Normal   ○ Large
1) $\frac{\pi}{6}$
2) $\frac{\pi}{3}$
3) $-\frac{\pi}{6}$
4) $-\frac{\pi}{3}$

7.14. Calculate the value of the definite integral below.

$$\int_0^1 \frac{1}{\sqrt{2x - x^2}} dx$$

Difficulty level   ○ Easy   ● Normal   ○ Hard
Calculation amount   ● Small   ○ Normal   ○ Large
1) 0
2) $\frac{\pi}{4}$
3) $\frac{\pi}{2}$
4) π

7.15. Calculate the value of the definite integral below.

$$\int_{-1}^{1} (x^2 + 1)(x^3 + 3x)\,dx$$

Difficulty level   ○ Easy   ● Normal   ○ Hard
Calculation amount   ● Small   ○ Normal   ○ Large
1) 21
2) 0
3) −11
4) 2

7.16. Solve the following indefinite integral.

$$\int \frac{\sin(x)}{1 + \cos(\cos(x))}\,dx$$

Difficulty level   ○ Easy   ● Normal   ○ Hard
Calculation amount   ○ Small   ● Normal   ○ Large
1) $-\tan\left(\frac{1}{2}\cos(x)\right) + c$
2) $\tan(\cos(x)) + c$
3) $-\tan(\cos(x)) + c$
4) $\tan\left(\frac{1}{2}\cos(x)\right) + c$

7.17. Calculate the surface area enclosed between the curves of $y = 2x^2 - 2x$ and $y = x^2$.

Difficulty level   ○ Easy   ● Normal   ○ Hard
Calculation amount   ○ Small   ● Normal   ○ Large
1) $\frac{1}{3}$
2) $\frac{2}{3}$
3) $\frac{4}{3}$
4) $\frac{7}{3}$

7.18. Determine the function of a curve that passes from the point of (3, 4) and its derivative is $-\frac{x}{y}$.

Difficulty level   ○ Easy   ● Normal   ○ Hard
Calculation amount   ○ Small   ● Normal   ○ Large
1) $2x^2 + y^2 = 34$
2) $x^2 + y^2 = 16$
3) $y^2 = 4x + 4$
4) $x^2 + y^2 = 25$

7.19. Solve the following indefinite integral.

$$\int \frac{x}{\sqrt{x-1}}\,dx$$

Difficulty level   ○ Easy   ● Normal   ○ Hard
Calculation amount   ○ Small   ● Normal   ○ Large

1) $\frac{2}{3}(x-1)^{\frac{3}{2}} - 2(x-1)^{\frac{1}{2}} + c$

2) $\frac{2}{3}(x-1)^{\frac{3}{2}} + 2(x-1)^{\frac{1}{2}} + c$

3) $\frac{1}{3}(x-1)^{\frac{3}{2}} - 2(x-1)^{\frac{1}{2}} + c$

4) $-\frac{1}{3}(x-1)^{\frac{3}{2}} - 2(x-1)^{\frac{1}{2}} + c$

7.20. Calculate the value of the definite integral below.

$$\int_{-2}^{5} |x-3| dx$$

Difficulty level   ○ Easy   ● Normal   ○ Hard
Calculation amount   ○ Small   ● Normal   ○ Large

1) $\frac{25}{2}$

2) $\frac{27}{2}$

3) $\frac{29}{2}$

4) $\frac{31}{2}$

7.21. Calculate the value of the definite integral below.

$$\int_{1}^{2} \frac{x(x+1)^2 + 2}{(x+1)^2} dx$$

Difficulty level   ○ Easy   ● Normal   ○ Hard
Calculation amount   ○ Small   ● Normal   ○ Large

1) $\frac{12}{5}$

2) $\frac{9}{5}$

3) $\frac{11}{6}$

4) $\frac{7}{4}$

7.22. Calculate the surface area enclosed between the curves of $y = x^2$ and $y = \sqrt{x}$.

Difficulty level   ○ Easy   ● Normal   ○ Hard
Calculation amount   ○ Small   ● Normal   ○ Large

1) $\frac{2}{3}$

2) 1

3) $\frac{1}{3}$

4) $\frac{1}{6}$

7.23. Solve the following indefinite integral.

$$\int \frac{1}{1+e^x} dx$$

Difficulty level   ○ Easy   ● Normal   ○ Hard
Calculation amount   ○ Small   ● Normal   ○ Large

1) $x + \ln(1+e^x) + c$

2) $x - \ln(1+e^x) + c$

3) $\frac{1}{2}x^2 + \ln(1+e^x) + c$

4) $\frac{1}{2}x^2 - \ln(1+e^x) + c$

7.24. Calculate the surface area enclosed between the curve of $y = x^3 + 2x^2 + x$ and x-axis.

Difficulty level   ○ Easy   ● Normal   ○ Hard
Calculation amount   ○ Small ● Normal   ○ Large

1) $\frac{1}{12}$

2) $\frac{1}{10}$

3) $\frac{1}{9}$

4) $\frac{1}{7}$

7.25. Calculate the mean value of the function of $y = ax + b$ in the range of [2, 5].

Difficulty level   ○ Easy   ● Normal   ○ Hard
Calculation amount   ○ Small ● Normal   ○ Large

1) $\frac{5}{2}a + 3b$

2) $\frac{7}{2}a + 3b$

3) $\frac{5}{2}a + b$

4) $\frac{7}{2}a + b$

7.26. Determine the function of a curve that passes from the point of (1, 1) and its derivative is as follows.

$$y' = \frac{x+1}{1-y}$$

Difficulty level   ○ Easy   ● Normal   ○ Hard
Calculation amount   ○ Small ● Normal   ○ Large

1) $x^2 + y^2 + 2x - 2y - 2 = 0$
2) $x^2 - y^2 + 4x - 4y + 1 = 0$
3) $x^2 + y^2 - 2x + 2y - 2 = 0$
4) $x^2 - y^2 + 3x - 2y - 1 = 0$

7.27. Consider the functions of $f(x) = 2x$ and $g(x) = 3x^2 - 2x$. Calculate the value of λ if the mean value of the functions in the range of $[1, \lambda]$ is the same.

Difficulty level   ○ Easy   ● Normal   ○ Hard
Calculation amount   ○ Small ● Normal   ○ Large

1) $\frac{1+\sqrt{5}}{2}$

2) $\frac{1+\sqrt{3}}{2}$

3) $\frac{\sqrt{5}}{2}$

4) $\frac{3\sqrt{3}}{2}$

7.28. What is the function of a curve that passes from the point of (1, 1) and the relation below holds.

$$y' = \frac{3x}{2y}$$

Difficulty level   ○ Easy   ● Normal   ○ Hard
Calculation amount   ○ Small ● Normal   ○ Large

1) $2y^2 - 3x^2 + 1 = 0$
2) $y^2 - 2x^2 + 1 = 0$
3) $2y^2 + x^2 - 3 = 0$
4) $2y^2 - x^2 - 1 = 0$

7.29. In the equation below, determine the value of A.

$$\int \frac{3x}{\sqrt{x^2 + 1}} dx = A\sqrt{x^2 + 1} + c$$

Difficulty level   ○ Easy   ● Normal   ○ Hard
Calculation amount   ○ Small   ● Normal   ○ Large

1) $\frac{1}{2}$
2) 1
3) $\frac{3}{2}$
4) 3

7.30. Calculate the value of the definite integral below.

$$\int_{-1}^{2} |x| dx$$

Difficulty level   ○ Easy   ● Normal   ○ Hard
Calculation amount   ○ Small   ● Normal   ○ Large

1) $\frac{3}{2}$
2) $\frac{5}{2}$
3) $\frac{7}{2}$
4) $\frac{9}{2}$

7.31. Calculate the value of the following definite integral.

$$\int_{0}^{\frac{\pi}{2}} \sin^2(x) dx$$

Difficulty level   ○ Easy   ● Normal   ○ Hard
Calculation amount   ○ Small   ● Normal   ○ Large

1) 0
2) $\frac{\pi}{2}$
3) $\frac{\pi}{4}$
4) $\frac{\pi}{8}$

7.32. Calculate the value of the definite integral below.

$$\int_{0}^{\frac{3\pi}{4}} \left(\tan^5(x) + \tan^7(x) \right) dx$$

Difficulty level   ○ Easy   ● Normal   ○ Hard
Calculation amount   ○ Small   ● Normal   ○ Large

1) 0
2) $\frac{1}{2}$
3) $\frac{1}{3}$
4) $\frac{1}{6}$

7.33. Which one of the points below is on a curve that passes from the point of $(\pi, 1)$ and $y' = y^2 \cos(x)$ holds for that?

Difficulty level   ○ Easy   ● Normal   ○ Hard
Calculation amount   ○ Small   ● Normal   ○ Large

1) $\left(\frac{3\pi}{2}, 2\right)$
2) $\left(\frac{\pi}{2}, -1\right)$
3) $\left(\frac{\pi}{2}, 1\right)$
4) $(0, 1)$

7.34. Calculate the value of the definite integral of I_1 if $I_2 = m$.

$$I_1 = \int_3^5 \frac{3x}{x-2} dx$$

$$I_2 = \int_3^5 \frac{1}{x-2} dx$$

Difficulty level   ○ Easy   ● Normal   ○ Hard
Calculation amount   ○ Small   ● Normal   ○ Large

1) $m + 2$
2) $4m - 6$
3) $6m + 6$
4) $6m - 4$

7.35. Calculate the value of the definite integral below.

$$\int_2^3 \frac{x}{x^2 - 1} dx$$

Difficulty level   ○ Easy   ● Normal   ○ Hard
Calculation amount   ○ Small   ● Normal   ○ Large

1) $\ln\left(\frac{8}{3}\right)$
2) $\mathrm{arc}\left(\sin\left(\frac{2\sqrt{3}}{5}\right)\right)$
3) $\mathrm{arc}\left(\tan\left(\frac{3}{2}\right)\right)$
4) $\ln\left(\sqrt{\frac{8}{3}}\right)$

7.36. Calculate the volume resulted from the rotation of the surface area around x-axis enclosed between one period of the curve of $y = \sin(x)$ and x-axis.

Difficulty level   ○ Easy   ● Normal   ○ Hard
Calculation amount   ○ Small   ● Normal   ○ Large

1) π^2
2) $2\pi^2$
3) $\frac{\pi^2}{2}$
4) $\frac{\pi^2}{4}$

7.37. Calculate the value of the definite integral of $\int_0^4 f'(x)dx$ if we have $f(x) = \int_a^x \sqrt{t}\,dt$.

Difficulty level   ○ Easy   ● Normal   ○ Hard
Calculation amount   ○ Small   ● Normal   ○ Large

1) $\frac{8}{3}$

2) $\frac{16}{3}$

3) $-\frac{8}{3}$

4) $-\frac{16}{3}$

7.38. Solve the following indefinite integral.

$$\int 8\left(\tan^6(x) + \tan^8(x)\right)dx$$

Difficulty level   ○ Easy   ● Normal   ○ Hard

Calculation amount   ○ Small   ● Normal   ○ Large

1) $\tan^7(x) + c$

2) $\frac{1}{7}\tan^8(x) + c$

3) $\frac{8}{5}\tan^5(x) + c$

4) $\frac{8}{7}\tan^7(x) + c$

7.39. Calculate the value of the definite integral below.

$$\int_{-1}^{1}(x+1)\left(x^2 + 2x + 3\right)dx$$

Difficulty level   ○ Easy   ● Normal   ○ Hard

Calculation amount   ○ Small   ● Normal   ○ Large

1) 6

2) 4

3) 8

4) 10

7.40. Calculate the value of the definite integral below.

$$\int_{\frac{1}{2}}^{1}\left[\frac{1}{x}\right]\frac{1}{x^3}dx$$

Difficulty level   ○ Easy   ● Normal   ○ Hard

Calculation amount   ○ Small   ● Normal   ○ Large

1) 1

2) 2

3) $\frac{1}{2}$

4) $\frac{3}{2}$

7.41. Calculate the value of y'_x if we have:

$$y = u + v, \quad u = \int_{1}^{x^2}\frac{\sin(t)}{t}dt, \quad v = \int_{x^2}^{1}\frac{\sin(u)}{u}du$$

Difficulty level   ○ Easy   ● Normal   ○ Hard

Calculation amount   ○ Small   ● Normal   ○ Large

1) $\dfrac{4\sin\left(x^2\right)}{x}$

2) $-\dfrac{4\sin\left(x^2\right)}{x}$

3) 0

4) 1

7.42. Calculate the surface area enclosed between the curve of $y = x^2 + 1$ and the line of $y = 2$.

Difficulty level   ○ Easy   ● Normal   ○ Hard
Calculation amount   ○ Small   ● Normal   ○ Large

1) $\frac{1}{3}$

2) $\frac{2}{3}$

3) 1

4) $\frac{4}{3}$

7.43. Calculate the value of $f(3x + 2)$ if we have:

$$h(x) = \int f'(3x + 2)dx, \quad h(0) = 1, \quad f(2) = 3$$

Difficulty level   ○ Easy   ● Normal   ○ Hard
Calculation amount   ○ Small   ● Normal   ○ Large

1) $h(x) - 1$

2) $2h(x) + 1$

3) $3h(x)$

4) $3h(x) - 1$

7.44. Calculate the surface area enclosed between the curve with the function below and x-axis in the range of $\left[-\sqrt{2}, \sqrt{2}\right]$.

$$y = \frac{1}{2 + x^2}$$

Difficulty level   ○ Easy   ● Normal   ○ Hard
Calculation amount   ○ Small   ● Normal   ○ Large

1) $\frac{\pi\sqrt{2}}{4}$

2) $\frac{\pi}{4}$

3) $\frac{\pi\sqrt{2}}{2}$

4) $\frac{\pi}{2}$

7.45. A curve is tangent to $y = x$ in the origin and its second derivative is $2x + 1$. Which one of the points below is on the curve?

Difficulty level   ○ Easy   ● Normal   ○ Hard
Calculation amount   ○ Small   ○ Normal   ● Large

1) $\left(1, \frac{11}{6}\right)$

2) $\left(1, \frac{13}{6}\right)$

3) $\left(2, \frac{11}{6}\right)$

4) $\left(2, \frac{13}{6}\right)$

7.46. Solve the indefinite integral of $\int \sin (2x) \cos (4x) dx$.

Difficulty level   ○ Easy   ● Normal   ○ Hard
Calculation amount   ○ Small   ○ Normal   ● Large
1) $\frac{1}{3} \cos^3 (2x) - \frac{1}{2} \cos (2x) + c$
2) $-\frac{1}{3} \cos^3 (2x) + \frac{1}{2} \cos (2x) + c$
3) $\frac{1}{3} \cos^3 (2x) + \frac{1}{2} \cos (2x) + c$
4) $-\frac{1}{3} \cos^3 (2x) - \frac{1}{2} \cos (2x) + c$

7.47. Calculate the value of the definite integral below.

$$\int_{-1}^{1} \left[\frac{x}{3} \right] dx$$

Difficulty level   ○ Easy   ○ Normal   ● Hard
Calculation amount   ● Small   ○ Normal   ○ Large
1) 0
2) 1
3) −1
4) −3

7.48. What is the function of a curve that passes from the point of (1, 2) and the relation of $xy' + y = 1$ holds.

Difficulty level   ○ Easy   ○ Normal   ● Hard
Calculation amount   ● Small   ○ Normal   ○ Large
1) $y = 1 + \frac{1}{x}$
2) $y = 2 - \frac{1}{x}$
3) $y = \frac{3}{x} - 1$
4) $y = \frac{3}{x} + 1$

7.49. Solve the indefinite integral below.

$$\int \frac{f'(\sqrt[3]{x})}{\sqrt[3]{x^2}} dx$$

Difficulty level   ○ Easy   ○ Normal   ● Hard
Calculation amount   ● Small   ○ Normal   ○ Large
1) $\frac{1}{3} f(\sqrt[3]{x}) + c$
2) $\frac{2}{3} f(\sqrt[3]{x}) + c$
3) $f(\sqrt[3]{x}) + c$
4) $3 f(\sqrt[3]{x}) + c$

7.50. Solve the indefinite integral below.

$$\int \ln x dx$$

Difficulty level   ○ Easy   ○ Normal   ● Hard
Calculation amount   ○ Small   ● Normal   ○ Large

1) $x \ln x - x + c$
2) $x \ln x + x + c$
3) $-x \ln x + x + c$
4) $-x \ln x - x + c$

7.51. Solve the following indefinite integral.

$$\int \frac{1}{\sin(x)\cos(x)} dx$$

Difficulty level   ○ Easy   ○ Normal   ● Hard
Calculation amount   ○ Small   ● Normal   ○ Large
1) $\ln|\sin(2x)| + c$
2) $\ln|\tan(x)| + c$
3) $\ln|\cos(2x)| + c$
4) $\ln|\cot(x)| + c$

7.52. Calculate the value of the definite integral below.

$$\int_0^{\frac{\pi}{4}} \frac{1}{\cos^4(x)} dx$$

Difficulty level   ○ Easy   ○ Normal   ● Hard
Calculation amount   ○ Small   ● Normal   ○ Large
1) $\frac{1}{3}$
2) $\frac{2}{3}$
3) 1
4) $\frac{4}{3}$

7.53. Calculate the value of the definite integral below.

$$\int_0^{\frac{\pi}{4}} \frac{1}{\sqrt[3]{\sin^2(x)\cos^4(x)}} dx$$

Difficulty level   ○ Easy   ○ Normal   ● Hard
Calculation amount   ○ Small   ● Normal   ○ Large
1) 1
2) 2
3) 3
4) 4

7.54. Calculate the value of $f(x = e)$ if the derivative of $f(x^2)$ with respect to x is $\frac{6}{x}$ and $f(x = 1) = 0$.

Difficulty level   ○ Easy   ○ Normal   ● Hard
Calculation amount   ○ Small   ● Normal   ○ Large
1) 0
2) 1
3) 3
4) 6

7.55. Calculate the value of $f(x = -1)$ if $f'(cos^2(x)) = \cos(2x)$ and $f(x = 1) = 1$.

1) 1
2) 2
3) 3
4) 4

7.56. Solve the following indefinite integral.

$$\int \frac{\cos(2x)}{\sin^2(x)\cos^2(x)}dx$$

1) $-\frac{2}{\sin(2x)} + c$
2) $-\frac{1}{\sin(2x)} + c$
3) $\frac{2}{\sin(2x)} + c$
4) $\frac{1}{\sin(2x)} + c$

7.57. Calculate the value of the definite integral below.

$$\int_1^e (2x + \ln(x))dx$$

1) e^2
2) 1
3) $1 + e$
4) $e - 1$

7.58. Solve the indefinite integral below.

$$\int \sin(2x)(2 + \cos^2(x))^{50}dx$$

1) $\frac{1}{51}(2 + \cos^2(x))^{51} + c$
2) $-\frac{1}{51}(2 + \cos^2(x))^{51} + c$
3) $\frac{1}{51}(2 + \cos^2(x))^{50} + c$
4) $-\frac{1}{51}(2 + \cos^2(x))^{50} + c$

7.59. Calculate the value of the definite integral below.

$$\int_1^e \frac{\ln(x)}{x}\,dx$$

Difficulty level   ○ Easy   ○ Normal   ● Hard
Calculation amount   ○ Small   ● Normal   ○ Large
1) 1
2) $\frac{1}{2}$
3) 2
4) $\frac{1}{e}$

7.60. Determine the function of a curve that passes from the point of (0, 1) and the relation below holds.

$$y' = -\frac{2x+2}{4y+1}$$

Difficulty level   ○ Easy   ○ Normal   ● Hard
Calculation amount   ○ Small   ● Normal   ○ Large
1) $2x^2 + y^2 = 34 + 3x$
2) $x^2 - y^2 = -7y + 5$
3) $x^2 + y^2 = 4x + 4y - 1$
4) $x^2 + 2y^2 = -y - 2x + 3$

7.61. Calculate the value of the definite integral below.

$$\int_{\frac{\pi}{6}}^{\frac{\pi}{2}} \frac{\cot(x)}{\sqrt{1-\cos(2x)}}\,dx$$

Difficulty level   ○ Easy   ○ Normal   ● Hard
Calculation amount   ○ Small   ● Normal   ○ Large
1) $\sqrt{2}$
2) $\frac{\sqrt{2}}{2}$
3) $\sqrt{3}$
4) $\frac{\sqrt{3}}{2}$

7.62. Calculate the value of the definite integral below.

$$\int_3^6 \frac{x+2}{\sqrt{x-2}}\,dx$$

Difficulty level   ○ Easy   ○ Normal   ● Hard
Calculation amount   ○ Small   ● Normal   ○ Large
1) $\frac{25}{3}$
2) $\frac{38}{3}$
3) $\frac{23}{3}$
4) $\frac{34}{3}$

7.63. Calculate the value of the definite integral below.

$$\int_1^4 \frac{\sqrt{1+\sqrt{x}}}{\sqrt{x}}\,dx$$

Difficulty level   ○ Easy   ○ Normal   ● Hard
Calculation amount   ○ Small   ● Normal   ○ Large

1) $2\left(\sqrt{3}+\frac{1}{3}\right)$
2) $2\left(\sqrt{3}-\frac{1}{3}\right)$
3) $4\left(\sqrt{3}+\frac{2\sqrt{2}}{3}\right)$
4) $4\left(\sqrt{3}-\frac{2\sqrt{2}}{3}\right)$

7.64. Solve the following indefinite integral if we know that $0 < x < \pi$.

$$\int \cot(x)\sqrt{\sin(x)}\,dx$$

Difficulty level   ○ Easy   ○ Normal   ● Hard
Calculation amount   ○ Small   ● Normal   ○ Large

1) $\sqrt{\cot(x)}+c$
2) $2\sqrt{\sin(x)}+c$
3) $\sin(x)\sqrt{\sin(x)}+c$
4) $\frac{1}{2}\sqrt{\sin(x)}+c$

7.65. Calculate the volume resulted from the rotation of the surface area around y-axis enclosed between the curve of $y = 1 - \frac{1}{4}x^2$ and x-axis.

Difficulty level   ○ Easy   ○ Normal   ● Hard
Calculation amount   ○ Small   ● Normal   ○ Large

1) π
2) 2π
3) 3π
4) 4π

7.66. Calculate the value of the definite integral below.

$$\int_0^{\frac{\pi}{3}} \sec(x)\tan(x)\,dx$$

Difficulty level   ○ Easy   ○ Normal   ● Hard
Calculation amount   ○ Small   ● Normal   ○ Large

1) 1
2) 2
3) $\frac{1}{2}$
4) $\frac{3}{2}$

7.67. Calculate the value of the definite integral below.

$$\int_{\frac{\pi}{6}}^{\frac{\pi}{4}} \csc(x)\cot(x)dx$$

Difficulty level   ○ Easy   ○ Normal   ● Hard
Calculation amount   ○ Small   ● Normal   ○ Large
1) $2 + \sqrt{2}$
2) $2 - \sqrt{2}$
3) $\sqrt{3} - \sqrt{2}$
4) $\sqrt{3} + \sqrt{2}$

7.68. Calculate the value of the definite integral below.

$$\int_{\frac{\pi}{6}}^{\frac{\pi}{4}} \frac{1}{\sin^2(x)\cos^2(x)}dx$$

Difficulty level   ○ Easy   ○ Normal   ● Hard
Calculation amount   ○ Small   ○ Normal   ● Large
1) $\frac{\sqrt{3}}{3}$
2) $\frac{2\sqrt{3}}{3}$
3) $\sqrt{3}$
4) $2\sqrt{3}$

7.69. Solve the following indefinite integral.

$$\int (\tan(x) - \cot(x))(\tan(x) + \cot(x))^5 dx$$

Difficulty level   ○ Easy   ○ Normal   ● Hard
Calculation amount   ○ Small   ○ Normal   ● Large
1) $\frac{1}{4}(\tan(x) + \cot(x))^4 + c$
2) $\frac{1}{5}(\tan(x) + \cot(x))^5 + c$
3) $\frac{1}{3}(\tan(x) + \cot(x))^3 + c$
4) $\frac{1}{5}(\tan(x) - \cot(x))^5 + c$

7.70. Calculate the volume resulted from the rotation of the surface area around x-axis enclosed between the curve with the function below, x-axis, $x = \frac{\pi}{4}$, and $x = \frac{\pi}{2}$.

$$y = \frac{1}{\sin^2(x)}$$

Difficulty level   ○ Easy   ○ Normal   ● Hard
Calculation amount   ○ Small   ○ Normal   ● Large
1) π
2) $\frac{2\pi}{3}$
3) 2π
4) $\frac{4\pi}{3}$

7.71. Which one of the choices is not an acceptable solution for the indefinite integral of $\int \sin(x) \cos(x) dx$.

    Difficulty level   ○ Easy   ○ Normal   ● Hard

    Calculation amount   ○ Small   ○ Normal   ● Large

    1) $-\frac{1}{4}\cos(2x) + c$

    2) $-\frac{1}{4}\sin(2x) + c$

    3) $-\frac{1}{2}\cos^2(x) + c$

    4) $\frac{1}{2}\sin^2(x) + c$

Reference

1. Rahmani-Andebili, M. (2020). Precalculus: Practice problems, methods, and solutions, Springer Nature, 2020.

Abstract

In this chapter, the problems of the seventh chapter are fully solved, in detail, step-by-step, and with different methods. The subjects include definite integrals, indefinite integrals, substitution rule for integrals, integration techniques, integration by parts, integrals involving trigonometric functions, trigonometric substitutions, integration using partial fractions, integrals involving roots, integrals involving quadratics, applications of integrals, average value, area between curves, and volume of solid of revolution.

8.1. From the list of integral of trigonometric functions, we know that [1]:

$$\int x^n dx = \frac{1}{n+1} x^{n+1} + c$$

The problem can be solved as follows.

$$\int_1^2 \frac{x+4}{x^3} dx = \int_1^2 \left(\frac{1}{x^2} + \frac{4}{x^3}\right) dx = \left(-\frac{1}{x} - \frac{2}{x^2}\right)\Big|_1^2 = -\frac{1}{2} - \frac{2}{4} - (-1 - 2) = 2$$

Choice (2) is the answer.

8.2. From the list of integral of trigonometric functions, we know that:

$$\int e^u du = e^u$$

The problem can be solved as follows.

$$\int \left(e^x + 2xe^{x^2}\right) dx = e^x + e^{x^2} + c$$

Choice (4) is the answer.

8.3. Based on the information given in the problem, we have:

$$f(x) = \int (x^3 + 5x) dx$$

$$\stackrel{\frac{d}{dx}}{\Rightarrow} f'(x) = x^3 + 5x \stackrel{\frac{d}{dx}}{\Rightarrow} f''(x) = 3x^2 + 5 \stackrel{x=1}{\Rightarrow} f''(1) = 3 + 5 = 8$$

Choice (4) is the answer.

8.4. The problem can be solved as follows.

$$f(x) = \int \cos^3(x)dx \Rightarrow f'(x) = \cos^3(x) \Rightarrow f''(x) = -3\cos^2(x)\sin(x)$$

$$f''\left(\frac{\pi}{2}\right) = -3 \times 0 \times 1 = 0$$

Choice (1) is the answer.

8.5. From the list of integral of trigonometric functions, we know that:

$$\int x^n dx = \frac{1}{n+1}x^{n+1} + c$$

The problem can be solved as follows.

$$\int_{-2}^{-1} \frac{x^3 + x^2 - 1}{x^2} dx = \int_{-2}^{-1}\left(x + 1 - \frac{1}{x^2}\right)dx = \left(\frac{x^2}{2} + x + \frac{1}{x}\right)\Big|_{-2}^{-1}$$

$$= \left(\frac{1}{2} - 1 - 1\right) - \left(2 - 2 - \frac{1}{2}\right) = -1$$

Choice (2) is the answer.

8.6. From the list of integral of trigonometric functions, we know that:

$$\int x^n dx = \frac{1}{n+1}x^{n+1} + c$$

The problem can be solved as follows.

$$\int \frac{x-2}{\sqrt{x}} dx = \int \left(x^{\frac{1}{2}} - 2x^{-\frac{1}{2}}\right)dx = \frac{2}{3}x^{\frac{3}{2}} - 4x^{\frac{1}{2}} + c = \frac{2}{3}\sqrt{x}(x-6) + c$$

Choice (2) is the answer.

8.7. The problem can be solved by changing the variable of the integral as follows.

$$x^3 + 1 \triangleq u \stackrel{\frac{d}{dx}}{\Rightarrow} 3x^2 dx = du \Rightarrow x^2 dx = \frac{du}{3}$$

$$\int_0^1 \frac{x^2}{(x^3+1)^4} dx = \int_{u_1}^{u_2} u^{-4}\frac{du}{3} = \frac{1}{3}\frac{u^{-3}}{-3}\Big|_{u_1}^{u_2} = -\frac{1}{9}(x^3+1)^{-3}\Big|_0^1$$

$$-\frac{1}{9}(1+1)^{-3} + \frac{1}{9}(0+1)^{-3} = -\frac{1}{72} + \frac{1}{9} = \frac{-1+8}{72} = \frac{7}{72}$$

Choice (4) is the answer.

8.8. From the list of integral of trigonometric functions, we know that:

$$\int \frac{1}{x^2 + a^2} dx = \frac{1}{a} \text{arc} \left(\tan \frac{x}{a} \right) + c$$

Therefore,

$$\int_{-\infty}^{+\infty} \frac{1}{x^2 + 4} dx = \frac{1}{2} \text{arc} \left(\tan \left(\frac{x}{2} \right) \right) \Big|_{-\infty}^{+\infty} = \frac{1}{2} \left(\frac{\pi}{2} - \left(-\frac{\pi}{2} \right) \right) = \frac{\pi}{2}$$

Choice (1) is the answer.

8.9. From the list of integral of trigonometric functions, we know that:

$$\int x^n dx = \frac{1}{n+1} x^{n+1} + c$$

The problem can be solved as follows.

$$\int_1^4 \frac{x - \sqrt{x}}{\sqrt{x}} dx = \int_1^4 \left(x^{\frac{1}{2}} - 1 \right) dx = \left(\frac{2}{3} x^{\frac{3}{2}} - x \right) \Big|_1^4 = \left(\frac{16}{3} - 4 \right) - \left(\frac{2}{3} - 1 \right) = \frac{5}{3}$$

Choice (4) is the answer.

8.10. Based on the information given in the problem, we have:

$$\int f(x) dx = \frac{x^3}{6} \overset{\frac{d}{dx}}{\Rightarrow} f(x) = \frac{x^2}{2}$$

Therefore,

$$\Rightarrow f \left(\frac{1}{x} \right) = \frac{\left(\frac{1}{x} \right)^2}{2} = \frac{1}{2x^2}$$

$$\Rightarrow \frac{d}{dx} \left(f \left(\frac{1}{x} \right) \right) = \frac{d}{dx} \left(\frac{1}{2x^2} \right) = \frac{-4x}{4x^4} = \frac{-1}{x^3}$$

Choice (1) is the answer.

8.11. As we know:

$$F(x) = \int_{v(x)}^{u(x)} f(x) dx \Rightarrow F'(x) = u'(x) F(u(x)) - v'(x) F(v(x))$$

The problem can be solved as follows.

$$F(\lambda) = \int_0^\lambda \frac{1}{x^4 + 2} dx \Rightarrow F'(\lambda) = 1 \times \frac{1}{\lambda^4 + 2} - 0 = \frac{1}{\lambda^4 + 2} \Rightarrow F'(0) = \frac{1}{2}$$

Choice (2) is the answer.

8.12. Since the function is an odd function and the range of the integral is symmetric, the final answer is zero.

$$\int_{-1}^{1} \frac{x^2}{1+x^2} \text{arc}(\tan(x)) dx = 0$$

Choice (1) is the answer.

8.13. From the list of integral of trigonometric functions, we know that:

$$\int \frac{1}{\sqrt{1-x^2}} dx = \text{arc}(\sin(x))$$

Therefore:

$$\int_{-\frac{1}{2}}^{\frac{1}{2}} \frac{1}{\sqrt{1-x^2}} dx = \text{arc}(\sin(x)) \Big|_{-\frac{1}{2}}^{\frac{1}{2}} = \frac{\pi}{6} - \left(-\frac{\pi}{6}\right) = \frac{\pi}{3}$$

Choice (2) is the answer.

8.14. From the list of integral of trigonometric functions, we know that:

$$\int \frac{1}{\sqrt{1-u^2}} dx = \text{arc}(\sin(u))$$

The problem can be solved by changing the variable of the integral as follows.

$$x - 1 \triangleq u \overset{\frac{d}{dx}}{\Rightarrow} dx = du$$

$$\int_0^1 \frac{1}{\sqrt{2x-x^2}} dx = \int_0^1 \frac{1}{\sqrt{1-(x-1)^2}} dx = \int_{u_1}^{u_2} \frac{1}{\sqrt{1-u^2}} du = (\text{arc}(\sin(u))) \Big|_{u_1}^{u_2}$$

$$= (\text{arc}(\sin(x-1))) \Big|_0^1 = 0 - \left(-\frac{\pi}{2}\right) = \frac{\pi}{2}$$

Choice (3) is the answer.

8.15. The final answer is zero, since the function is an odd function and the range of the integral is symmetric.

$$\int_{-1}^{1} (x^2+1)(x^3+3x) dx = 0$$

Choice (2) is the answer.

8.16. From trigonometry, we know that:

$$1 + \cos(u) = 2\cos^2\left(\frac{u}{2}\right)$$

$$\frac{1}{\cos^2\left(\frac{u}{2}\right)} = 1 + \tan^2\left(\frac{u}{2}\right)$$

In addition, from the list of integral of trigonometric functions, we know that:

$$\int\left(1+\tan^2\left(\frac{u}{a}\right)\right)du = a\tan\left(\frac{u}{a}\right)+c$$

The problem can be solved by changing the variable of the integral as follows.

$$\cos(x)\triangleq u \Rightarrow \frac{d}{dx}\cos(x)=\frac{d}{dx}u \Rightarrow -\sin(x)dx=du$$

$$\Rightarrow \int\frac{\sin(x)}{1+\cos(\cos(x))}dx = -\int\frac{1}{1+\cos(u)}du = -\int\frac{1}{2\cos^2\left(\frac{u}{2}\right)}du$$

$$= -\frac{1}{2}\int\left(1+\tan^2\left(\frac{u}{2}\right)\right)du = -\tan\left(\frac{u}{2}\right)+c = -\tan\left(\frac{\cos(x)}{2}\right)+c$$

Choice (1) is the answer.

8.17. From the list of integral of trigonometric functions, we know that:

$$\int x^n dx = \frac{1}{n+1}x^{n+1}+c$$

First, we need to find the intersection points of the curves as follows:

$$2x^2-2x=x^2 \Rightarrow x^2-2x=0 \Rightarrow x=0,2$$

$$S=\int_{x_1}^{x_2}(y_2-y_1)dx = \int_0^2(x^2-2x^2+2x)dx = \int_0^2(-x^2+2x)dx = \left(-\frac{x^3}{3}+x^2\right)\Big|_0^2$$

$$=\left(-\frac{2^3}{3}+2^2\right)-(0+0) \Rightarrow S=\frac{4}{3}$$

Choice (3) is the answer.

8.18. From the list of integral of trigonometric functions, we know that:

$$\int x^n dx = \frac{1}{n+1}x^{n+1}+c$$

Based on the information given in the problem, we have:

$$y(3)=4$$

$$y'=-\frac{x}{y}$$

The problem can be solved as follows.

$$y'=-\frac{x}{y} \Rightarrow yy'+x=0 \xrightarrow{\int dx} \frac{y^2}{2}+\frac{x^2}{2}=c' \Rightarrow y^2+x^2=c \qquad (1)$$

$$\xrightarrow{y(3) = 4} 4^2 + 3^2 = c \Rightarrow c = 25$$

$$\xrightarrow{(1), (2)} y^2 + x^2 = 25 \tag{2}$$

Choice (4) is the answer.

8.19. From the list of integral of trigonometric functions, we know that:

$$\int x^n dx = \frac{1}{n+1} x^{n+1} + c$$

The problem can be solved as follows.

$$\int \frac{x}{\sqrt{x-1}} dx = \int \frac{x-1+1}{\sqrt{x-1}} dx = \int \left((x-1)^{\frac{1}{2}} + (x-1)^{-\frac{1}{2}} \right) dx = \frac{2}{3} (x-1)^{\frac{3}{2}} + 2(x-1)^{\frac{1}{2}} + c$$

Choice (2) is the answer.

8.20. From the list of integral of trigonometric functions, we know that:

$$\int x^n dx = \frac{1}{n+1} x^{n+1} + c$$

The problem can be solved as follows.

$$\int_{-2}^{5} |x-3| dx = \int_{-2}^{3} (3-x) dx + \int_{3}^{5} (x-3) dx = \left(3x - \frac{x^2}{2} \right) \Big|_{-2}^{3} + \left(\frac{x^2}{2} - 3x \right) \Big|_{3}^{5}$$

$$= \left(9 - \frac{9}{2} \right) - (-6 - 2) + \left(\frac{25}{2} - 15 \right) - \left(\frac{9}{2} - 9 \right) = \frac{9}{2} + 8 - \frac{5}{2} + \frac{9}{2} = \frac{9 + 16 - 5 + 9}{2} = \frac{29}{2}$$

Choice (3) is the answer.

8.21. From the list of integral of trigonometric functions, we know that:

$$\int x^n dx = \frac{1}{n+1} x^{n+1} + c$$

The problem can be solved as follows.

$$\int_{1}^{2} \frac{x(x+1)^2 + 2}{(x+1)^2} dx = \int_{1}^{2} \left(x + 2(x+1)^{-2} \right) dx = \left(\frac{x^2}{2} - \frac{2}{x+1} \right) \Big|_{1}^{2} = \left(2 - \frac{2}{3} \right) - \left(\frac{1}{2} - 1 \right) = \frac{4}{3} + \frac{1}{2} = \frac{11}{6}$$

Choice (3) is the answer.

8.22. From the list of integral of trigonometric functions, we know that:

$$\int x^n dx = \frac{1}{n+1} x^{n+1} + c$$

First, we need to find the intersection points of the curves as follows:

$$\begin{cases} y_1 = x^2 \\ y_2 = \sqrt{x} \end{cases} \Rightarrow y_2 = y_1 \Rightarrow \sqrt{x} = x^2 \Rightarrow \sqrt{x}(x\sqrt{x} - 1) = 0 \Rightarrow x = 0, 1$$

$$S = \int_{x_1}^{x_2} (y_2 - y_1)dx = \int_0^1 (\sqrt{x} - x^2)dx = \left(\frac{2}{3}x^{\frac{3}{2}} - \frac{x^3}{3}\right)\Big|_0^1 = \frac{2}{3} - \frac{1}{3} \Rightarrow S = \frac{1}{3}$$

Choice (3) is the answer.

8.23. From the list of integral of trigonometric functions, we know that:

$$\int x^n dx = \frac{1}{n+1}x^{n+1} + c$$

$$\int \frac{1}{u} du = \ln|u| + c$$

The problem can be solved as follows.

$$\int \frac{1}{1 + e^x} dx = \int \frac{1 + e^x - e^x}{1 + e^x} dx = \int \left(1 - \frac{e^x}{1 + e^x}\right) dx = \int 1 dx - \int \frac{e^x}{1 + e^x} dx$$

$$= x + c' - \int \frac{e^x}{1 + e^x} dx \qquad (1)$$

Now, we should change the variable of the integral as follows.

$$1 + e^x \triangleq u \Rightarrow e^x dx = du \qquad (2)$$

Solving (1) and (2):

$$x + c' - \int \frac{1}{u} du = x + c' - \ln|u| + c'' = x - \ln|1 + e^x| + c = x - \ln(1 + e^x) + c$$

Choice (2) is the answer.

8.24. From the list of integral of trigonometric functions, we know that:

$$\int x^n dx = \frac{1}{n+1}x^{n+1} + c$$

First, we need to find the intersection points of the curves as follows:

$$y_2 = y_1 \Rightarrow x^3 + 2x^2 + x = 0 \Rightarrow x(x^2 + 2x + 1) = x(x+1)^2 = 0 \Rightarrow x = 0, -1, -1$$

$$S = \int_{x_1}^{x_2} (y_2 - y_1)dx = \int_{-1}^0 (x^3 + 2x^2 + x)dx = \left(\frac{x^4}{4} + \frac{2}{3}x^3 + \frac{x^2}{2}\right)\Big|_{-1}^0 = 0 - \left(\frac{1}{4} - \frac{2}{3} + \frac{1}{2}\right)$$

$$= -\left(\frac{3 - 8 + 6}{12}\right) = -\frac{1}{12}$$

The surface area must be a positive quantity. Therefore,

$$S = \left|-\frac{1}{12}\right| = \frac{1}{12}$$

Choice (1) is the answer.

8.25. As we know, the average value of a function can be determined as follows:

$$\frac{1}{b-a}\int_a^b f(x)dx$$

In addition, from the list of integral of trigonometric functions, we know that:

$$\int x^n dx = \frac{1}{n+1}x^{n+1} + c$$

The problem can be solved as follows.

$$\frac{1}{5-2}\int_2^5 (ax+b)dx = \frac{1}{3}\left(\frac{a}{2}x^2 + bx\right)\Big|_2^5 = \frac{1}{3}\left(\frac{25a}{2} + 5b - \frac{4a}{2} - 2b\right)$$

$$\text{The average value} = \frac{7}{2}a + b$$

Choice (4) is the answer.

8.26. The problem can be solved as follows.

$$y' = \frac{x+1}{1-y} \Rightarrow y' - yy' = x+1 \xrightarrow{\int dx} y - \frac{y^2}{2} = \frac{x^2}{2} + x + c$$

$$\xrightarrow{(x,y)=(1,1)} 1 - \frac{1}{2} = \frac{1}{2} + 1 + c \Rightarrow c = -1$$

$$\Rightarrow y - \frac{y^2}{2} = \frac{x^2}{2} + x - 1 \Rightarrow x^2 + y^2 + 2x - 2y - 2 = 0$$

Choice (1) is the answer.

8.27. As we know, the average value of a function can be determined as follows:

$$\frac{1}{b-a}\int_a^b f(x)dx$$

Moreover, from the list of integral of trigonometric functions, we know that:

$$\int x^n dx = \frac{1}{n+1}x^{n+1} + c$$

The problem can be solved as follows.

$$\frac{1}{\lambda-1}\int_1^\lambda 2x\,dx = \frac{1}{\lambda-1}\int_1^\lambda (3x^2 - 2x)dx \Rightarrow x^2\Big|_1^\lambda = (x^3 - x^2)\Big|_1^\lambda$$

$$\Rightarrow \lambda^2 - 1 = (\lambda^3 - \lambda^2) - 0 \Rightarrow \lambda^3 - 2\lambda^2 + 1 = 0 \Rightarrow (\lambda-1)(\lambda^2 - \lambda - 1) = 0$$

$$\Rightarrow \lambda = \frac{1-\sqrt{5}}{2}, \frac{1+\sqrt{5}}{2}, 1$$

However, just $\frac{1+\sqrt{5}}{2}$ is acceptable because the others are not within the range.

$$\Rightarrow \lambda = \frac{1+\sqrt{5}}{2}$$

Choice (1) is the answer.

8.28. The problem can be solved as follows.

$$y' = \frac{3x}{2y} \Rightarrow 2yy' = 3x \Rightarrow y^2 = \frac{3}{2}x^2 + c$$

$$\xrightarrow{(x,y)=(1,1)} 1 = \frac{3}{2} + c \Rightarrow c = \frac{-1}{2}$$

$$\Rightarrow y^2 = \frac{3}{2}x^2 - \frac{1}{2} \Rightarrow 2y^2 - 3x^2 + 1 = 0$$

Choice (1) is the answer.

8.29. Based on the information given in the problem, we have:

$$\int \frac{3x}{\sqrt{x^2+1}} dx = A\sqrt{x^2+1} + c \qquad (1)$$

From the list of integral of trigonometric functions, we know that:

$$\int \frac{1}{\sqrt{u}} du = 2\sqrt{u} + c$$

The problem can be solved by changing the variable of the integral as follows.

$$x^2 + 1 \triangleq u \xrightarrow{\frac{d}{dx}} 2xdx = du \Rightarrow xdx = \frac{1}{2}du$$

$$\int \frac{3x}{\sqrt{x^2+1}} dx = \int \frac{3 \times \frac{1}{2}du}{\sqrt{u}} = \frac{3}{2}\int \frac{1}{\sqrt{u}} du = \frac{3}{2} \times 2\sqrt{u} = 3\sqrt{x^2+1} + c \qquad (2)$$

Therefore,

$$\xrightarrow{(1),(2)} A\sqrt{x^2+1} + c = 3\sqrt{x^2+1} + c \Rightarrow A = 3$$

Choice (4) is the answer.

8.30. From the list of integral of trigonometric functions, we know that:

$$\int x^n dx = \frac{1}{n+1}x^{n+1} + c$$

The problem can be solved as follows.

$$\int_{-1}^{2} |x| dx = \int_{-1}^{0} |x| dx + \int_{0}^{2} |x| dx = \int_{-1}^{0} (-x)dx + \int_{0}^{2} xdx = -\frac{x^2}{2}\Big|_{-1}^{0} + \frac{x^2}{2}\Big|_{0}^{2} = -\left(0 - \frac{1}{2}\right) + (2-0) = \frac{5}{2}$$

Choice (2) is the answer.

8.31. From trigonometry, we know that:

$$1 - \cos(2x) = 2\sin^2(x)$$

Furthermore, from the list of integral of trigonometric functions, we know that:

$$\int x^n dx = \frac{1}{n+1} x^{n+1} + c$$

$$\int \cos(ax) dx = \frac{1}{a} \sin(ax) + c$$

The problem can be solved as follows.

$$\int_0^{\frac{\pi}{2}} \sin^2(x) dx = \int_0^{\frac{\pi}{2}} \left(\frac{1}{2} - \frac{\cos(2x)}{2} \right) dx = \left(\frac{1}{2} x - \frac{1}{4} \sin(2x) \right) \Big|_0^{\frac{\pi}{2}} = \left(\frac{1}{2} \times \frac{\pi}{2} - 0 \right) - (0) = \frac{\pi}{4}$$

Choice (3) is the answer.

8.32. From the list of integral of trigonometric functions, we know that:

$$\int x^n dx = \frac{1}{n+1} x^{n+1} + c$$

The problem can be solved by changing the variable of the integral as follows.

$$\tan(x) \triangleq u \xrightarrow{\frac{d}{dx}} (1 + \tan^2(x)) dx = du$$

$$\int_0^{\frac{3\pi}{4}} \left(\tan^5(x) + \tan^7(x) \right) dx = \int_0^{\frac{3\pi}{4}} \tan^5(x) \left(1 + \tan^2(x) \right) dx$$

$$= \int_{u_1}^{u_2} u^5 du = \frac{1}{6} u^6 \Big|_{u_1}^{u_2} = \frac{1}{6} \tan^6(x) \Big|_0^{\frac{3\pi}{4}} = \frac{1}{6} (-1)^6 - 0 = \frac{1}{6}$$

Choice (4) is the answer.

8.33. The problem can be solved as follows.

$$y' = y^2 \cos(x) \Rightarrow \frac{y'}{y^2} = \cos(x) \Rightarrow \frac{-1}{y} = \sin(x) + c$$

$$\xrightarrow{(x,y) = (\pi, 1)} -1 = 0 + c \Rightarrow c = -1 \Rightarrow \frac{-1}{y} = \sin(x) - 1$$

We need to check each choice as follows:

$$\text{Choice 1}: \xrightarrow{(x,y) = \left(\frac{3\pi}{2}, 2 \right)} \frac{-1}{2} = \sin\left(\frac{3\pi}{2} \right) - 1 \Rightarrow \frac{-1}{2} \neq -2$$

$$\text{Choice 2}: \xrightarrow{(x,y) = \left(\frac{\pi}{2}, -1 \right)} \frac{-1}{-1} = \sin\left(\frac{\pi}{2} \right) - 1 \Rightarrow 1 \neq 0$$

$$\text{Choice 3}: \xrightarrow{(x,y)=\left(\frac{\pi}{2},1\right)} \frac{-1}{1} = \sin\left(\frac{\pi}{2}\right) - 1 \Rightarrow -1 \neq 0$$

$$\text{Choice 4}: \xrightarrow{(x,y)=(0,1)} \frac{-1}{1} = \sin(0) - 1 \Rightarrow -1 = -1$$

Choice (4) is the answer.

8.34. From the list of integral of trigonometric functions, we know that:

$$\int x^n dx = \frac{1}{n+1} x^{n+1} + c$$

Based on the information given in the problem, we have:

$$I_2 = \int_3^5 \frac{1}{x-2} dx = m \tag{1}$$

The problem can be solved as follows.

$$I_1 = \int_3^5 \frac{3x}{x-2} dx = \int_3^5 \frac{3x-6+6}{x-2} dx = \int_3^5 3dx + 6\int_3^5 \frac{1}{x-2} dx \tag{2}$$

Solving (1) and (2):

$$I_1 = \int_3^5 \frac{3x}{x-2} dx = 3x\Big|_3^5 + 6m = 15 - 9 + 6m = 6 + 6m$$

Choice (3) is the answer.

8.35. From the list of integral of trigonometric functions, we know that:

$$\int \frac{1}{u} du = \ln|u| + c$$

The problem can be solved as follows.

$$\int_2^3 \frac{x}{x^2-1} dx = \frac{1}{2} \int_2^3 \frac{2x}{x^2-1} dx \tag{1}$$

Now, we need to change the variable of the integral as follows.

$$x^2 - 1 \triangleq u \xrightarrow{\frac{d}{dx}} 2xdx = du \tag{2}$$

Solving (1) and (2):

$$\frac{1}{2} \int_{u_1}^{u_2} \frac{1}{u} du = \frac{1}{2} \ln|u|\Big|_{u_1}^{u_2} = \frac{1}{2} \ln|x^2-1|\Big|_2^3 = \frac{1}{2} \ln 8 - \frac{1}{2} \ln 3 = \frac{1}{2} \ln \frac{8}{3} = \ln \sqrt{\frac{8}{3}}$$

Choice (4) is the answer.

8.36. The volume resulted from the rotation of a surface area around x-axis, enclosed between the curve of $f(x)$ and x-axis, is calculated as follows:

$$V = \pi \int_{x_1}^{x_2} (f(x))^2 dx$$

Moreover, from trigonometry, we know that:

$$1 - \cos(2x) = 2\sin^2(x)$$

In addition, from the list of integral of trigonometric functions, we know that:

$$\int x^n dx = \frac{1}{n+1} x^{n+1} + c$$

$$\int \cos(ax) dx = \frac{1}{a} \sin(ax) + c$$

Therefore,

$$V = \pi \int_{0}^{2\pi} \sin^2(x) dx = \pi \int_{0}^{2\pi} \left(\frac{1}{2} - \frac{\cos(2x)}{2} \right) dx = \pi \left(\frac{x}{2} - \frac{1}{4} \sin(2x) \right) \Big|_{0}^{2\pi} = \pi(\pi - 0) = \pi^2$$

Choice (1) is the answer.

8.37. From the list of integral of trigonometric functions, we know that:

$$\int x^n dx = \frac{1}{n+1} x^{n+1} + c$$

As we know:

$$F(x) = \int_{v(x)}^{u(x)} f(x) dx \Longrightarrow F'(x) = u'(x) F(u(x)) - v'(x) F(v(x))$$

Therefore,

$$f(x) = \int_{a}^{x} \sqrt{t} \, dt \Longrightarrow f'(x) = \sqrt{x}$$

$$\int_{0}^{4} f'(x) dx = \int_{0}^{4} x^{\frac{1}{2}} dx = \left(\frac{2}{3} x^{\frac{3}{2}} \right) \Big|_{0}^{4} = \frac{16}{3}$$

Choice (2) is the answer.

8.38. From the list of integral of trigonometric functions, we know that:

$$\int x^n dx = \frac{1}{n+1} x^{n+1} + c$$

The problem can be solved by changing the variable of the integral as follows.

$$\tan(x) \triangleq u \overset{\frac{d}{dx}}{\Rightarrow} \left(1 + \tan^2(x)\right)dx = du$$

$$\int 8\left(\tan^6(x) + \tan^8(x)\right)dx = 8\int \tan^6(x)\left(1 + \tan^2(x)\right)dx$$

$$= 8\int u^6 du = \frac{8}{7}u^7 + c = \frac{8}{7}\tan^7(x) + c$$

Choice (4) is the answer.

8.39. From the list of integral of trigonometric functions, we know that:

$$\int x^n dx = \frac{1}{n+1}x^{n+1} + c$$

The problem can be solved as follows.

$$\int_{-1}^{1}(x+1)(x^2 + 2x + 3)dx = \int_{-1}^{1}(x+1)\left((x+1)^2 + 2\right)dx = \int_{-1}^{1}\left((x+1)^3 + 2(x+1)\right)dx \qquad (1)$$

Now, we should change the variable of the integral as follows.

$$x + 1 \triangleq u \Rightarrow dx = du \qquad (2)$$

Solving (1) and (2):

$$\int_{-1}^{1}\left(u^3 + 2u\right)dx = \frac{u^4}{4} + u^2 = \left(\frac{(x+1)^4}{4} + (x+1)^2\right)\Big|_{-1}^{1} = \frac{16}{4} + 4 - 0 = 8$$

Choice (3) is the answer.

8.40. From the list of integral of trigonometric functions, we know that:

$$\int x^n dx = \frac{1}{n+1}x^{n+1} + c$$

The problem can be solved as follows.

$$\int_{\frac{1}{2}}^{1}\left[\frac{1}{x}\right] \times \frac{1}{x^3}dx = \int_{\frac{1}{2}}^{1}1 \times x^{-3}dx = \left(\frac{x^{-2}}{-2}\right)\Big|_{\frac{1}{2}}^{1} = \left(\frac{-1}{2x^2}\right)\Big|_{\frac{1}{2}}^{1} = \frac{-1}{2} - (-2) = \frac{3}{2}$$

Choice (4) is the answer.

8.41. As we know:

$$F(x) = \int_{v(x)}^{u(x)}f(x)dx \Rightarrow F'(x) = u'(x)F(u(x)) - v'(x)F(v(x))$$

Therefore,

$$y'_x = u'_x + v'_x = \left(2x\frac{\sin(x^2)}{x^2} - 0\right) + \left(0 - 2x\frac{\sin(x^2)}{x^2}\right) = 0$$

Choice (3) is the answer.

8.42. From the list of integral of trigonometric functions, we know that:

$$\int x^n du = \frac{1}{n+1}x^{n+1} + c$$

First, we need to find the intersection points of the curves as follows:

$$\begin{cases} y_1 = x^2 + 1 \\ y_2 = 2 \end{cases} \Rightarrow x^2 + 1 = 2 \Rightarrow x^2 = 1 \Rightarrow x = \pm 1$$

$$S = \int_{-1}^{1}(2 - (x^2 + 1))dx = 2\int_{0}^{1}(1 - x^2)dx = 2\left(x - \frac{x^3}{3}\right)\Big|_{0}^{1} = 2\left(1 - \frac{1}{3}\right) - 0 = \frac{4}{3}$$

Choice (4) is the answer.

8.43. Based on the information given in the problem, we know that:

$$h(x) = \int f'(3x + 2)dx \qquad (1)$$

$$h(0) = 1 \qquad (2)$$

$$f(2) = 3 \qquad (3)$$

We should change the variable of the integral of $h(x)$ as follows.

$$f(3x + 2) \triangleq u \Rightarrow 3f'(3x + 2)dx = du \qquad (4)$$

Solving (1) and (4):

$$h(x) = \int \frac{1}{3}du = \frac{1}{3}u + c = \frac{1}{3}f(3x + 2) + c \qquad (5)$$

Solving (2) and (5):

$$1 = \frac{1}{3}f(2) + c \xrightarrow{\text{Using (3)}} 1 = \frac{1}{3} \times 3 + c \Rightarrow c = 0 \Rightarrow h(x) = \frac{1}{3}f(3x + 2) \Rightarrow f(3x + 2) = 3h(x)$$

Choice (3) is the answer.

8.44. From the list of integral of trigonometric functions, we know that:

$$\int \frac{1}{\left(1 + \left(\frac{x}{a}\right)^2\right)}dx = a \operatorname{arc}\left(\tan\left(\frac{x}{a}\right)\right)$$

The problem can be solved as follows.

$$S = \int_{-\sqrt{2}}^{\sqrt{2}} \frac{1}{2+x^2}\, dx = 2\int_{0}^{\sqrt{2}} \frac{1}{2+x^2}\, dx = 2\int_{0}^{\sqrt{2}} \frac{1}{2\left(1+\left(\frac{x}{\sqrt{2}}\right)^2\right)}\, dx$$

$$= \int_{0}^{\sqrt{2}} \frac{1}{\left(1+\left(\frac{x}{\sqrt{2}}\right)^2\right)}\, dx = \left(\sqrt{2}\, \mathrm{arc}\left(\tan\left(\frac{x}{\sqrt{2}}\right)\right)\right)\Big|_{0}^{\sqrt{2}} = \sqrt{2}\left(\frac{\pi}{4}-0\right) = \frac{\pi\sqrt{2}}{4}$$

Choice (1) is the answer.

8.45. From the list of integral of trigonometric functions, we know that:

$$\int x^n\, du = \frac{1}{n+1} x^{n+1} + c$$

Based on the information given in the problem, we know that:

$$y''(x) = 2x + 1 \tag{1}$$

The curve is tangent to $y = x$ in the origin. Thus:

$$y'(0) = 1 \tag{2}$$

$$y(0) = 0 \tag{3}$$

Applying integral operation on (1):

$$\int dx$$
$$\Longrightarrow y'(x) = x^2 + x + a \tag{4}$$

Applying integral operation on (4):

$$\int dx$$
$$\Longrightarrow y(x) = \frac{x^3}{3} + \frac{x^2}{2} + ax + b \tag{5}$$

Solving (2) and (4):

$$a = 1 \tag{6}$$

Solving (3) and (5):

$$0 = 0 + 0 + 0 + b \Longrightarrow b = 0 \tag{7}$$

Solving (5)–(7):

$$y(x) = \frac{x^3}{3} + \frac{x^2}{2} + x$$

Now, we need to check the choices as follows:

$$y(1) = \frac{1}{3} + \frac{1}{2} + 1 = \frac{2+3+6}{6} = \frac{11}{6}$$

$$y(2) = \frac{2^3}{3} + \frac{2^2}{2} + 2 = \frac{16+12+12}{6} = \frac{20}{3}$$

Choice (1) is the answer.

8.46. From trigonometry, we know that:

$$1 + \cos(2x) = 2\cos^2(x)$$

In addition, from the list of integral of trigonometric functions, we know that:

$$\int \sin(ax)dx = -\frac{1}{a}\cos(ax) + c$$

The problem can be solved as follows.

$$\int \sin(2x)\cos(4x)dx = \int \sin(2x)\left(2\cos^2(2x) - 1\right)dx$$

$$= \int \cos^2(2x) \times 2\sin(2x)dx - \int \sin(2x)dx \qquad (1)$$

Now, we should change the variable of the integral as follows.

$$\cos(2x) \triangleq u \implies -2\sin(2x)dx = du \qquad (2)$$

Solving (1) and (2):

$$-\int u^2 du - \int \sin(2x)dx = -\frac{1}{3}u^3 + \frac{1}{2}\cos(2x) + c = -\frac{1}{3}\cos^3(2x) + \frac{1}{2}\cos(2x) + c$$

Choice (2) is the answer.

8.47. From the list of integral of trigonometric functions, we know that:

$$\int x^n du = \frac{1}{n+1}x^{n+1} + c$$

The problem can be solved as follows.

$$\int_{-1}^{1} \left[\frac{x}{3}\right]dx = \int_{-1}^{0}(-1)dx + \int_{0}^{1}0dx = (-x)\Big|_{-1}^{0} + 0 = 0 - (1) = -1$$

Choice (3) is the answer.

8.48. The problem can be heuristically solved as follows.

$$xy' + y = 1 \Rightarrow (xy)' = 1 \Rightarrow xy = x + c$$

$$\xrightarrow[\quad]{(x,y) = (1,2)} 1 \times 2 = 1 + c \Rightarrow c = 1$$

$$\Rightarrow xy = x + 1 \Rightarrow y = 1 + \frac{1}{x}$$

Choice (1) is the answer.

8.49. The problem can be solved as follows.

$$\int \frac{f'(\sqrt[3]{x})}{\sqrt[3]{x^2}} dx = 3 \int \frac{1}{3\sqrt[3]{x^2}} f'(\sqrt[3]{x}) dx \qquad (1)$$

Now, we should change the variable of the integral as follows.

$$f(\sqrt[3]{x}) \triangleq u \Rightarrow \frac{1}{3\sqrt[3]{x^2}} f'(\sqrt[3]{x}) dx = du \qquad (2)$$

Solving (1) and (2):

$$3 \int du = 3u + c = 3f(\sqrt[3]{x}) + c$$

Choice (4) is the answer.

8.50. From integration by parts (partial integration), we know that:

$$\int u(x) dv = u(x)v(x) - \int v(x) du$$

In addition, from the list of integral of trigonometric functions, we know that:

$$\int \frac{1}{u} du = \ln|u| + c$$

The problem can be solved as follows.

$$\int \ln x dx \Rightarrow \begin{cases} u(x) = \ln x \\ dv - dx \end{cases} \Rightarrow \begin{cases} du = \frac{dx}{x} \\ v(x) = x \end{cases}$$

$$\Rightarrow \int \ln x dx = x \ln x - \int dx = x \ln x - x + c$$

Choice (1) is the answer.

8.51. From trigonometry, we know that:

$$\frac{1}{\cos^2(x)} = 1 + \tan^2(x)$$

$$\frac{\sin(x)}{\cos(x)} = \tan(x)$$

Moreover, from the list of integral of trigonometric functions, we know that:

$$\int \frac{1}{u} du = \ln|u| + c$$

The problem can be solved as follows.

$$\int \frac{1}{\sin(x)\cos(x)} dx = \int \frac{1}{\sin(x)\cos(x) \times \frac{\cos(x)}{\cos(x)}} dx = \int \frac{1}{\frac{\sin(x)}{\cos(x)}\cos^2(x)} dx = \int \frac{1 + \tan^2(x)}{\tan(x)} dx \qquad (1)$$

Now, we should change the variable of the integral as follows.

$$\tan(x) \triangleq u \implies \left(1 + \tan^2(x)\right) dx = du \qquad (2)$$

Solving (1) and (2):

$$\int \frac{1}{u} du = \ln|u| + c = \ln|\tan(x)| + c$$

Choice (2) is the answer.

8.52. From trigonometry, we know that:

$$1 + \tan^2(x) = \frac{1}{\cos^2(x)}$$

The problem can be solved as follows.

$$\int_0^{\frac{\pi}{4}} \frac{1}{\cos^4(x)} dx = \int_0^{\frac{\pi}{4}} \left(1 + \tan^2(x)\right)\left(1 + \tan^2(x)\right) dx \qquad (1)$$

Now, we should change the variable of the integral as follows.

$$\tan(x) \triangleq u \overset{\frac{d}{dx}}{\implies} \left(1 + \tan^2(x)\right) dx = du \qquad (2)$$

Solving (1) and (2):

$$\int_{u_1}^{u_2} \left(1 + u^2\right) du = \left(u + \frac{1}{3}u^3\right)\Big|_{u_1}^{u_2} = \left(\tan(x) + \frac{1}{3}\tan^3(x)\right)\Big|_0^{\frac{\pi}{4}} = \left(1 + \frac{1}{3}\right) - 0 = \frac{4}{3}$$

Choice (4) is the answer.

8.53. From trigonometry, we know that:

$$1 + \tan^2(x) = \frac{1}{\cos^2(x)}$$

$$\tan(x) = \frac{\sin(x)}{\cos(x)}$$

The problem can be solved as follows.

$$\int_0^{\frac{\pi}{4}} \frac{1}{\sqrt[3]{\sin^2(x)\cos^4(x)}} dx = \int_0^{\frac{\pi}{4}} \frac{1}{\sqrt[3]{\sin^2(x)\cos^4(x) \times \frac{\cos(x)}{\cos^2(x)}}} dx = \int_0^{\frac{\pi}{4}} \frac{\frac{1}{\cos^2(x)}}{\sqrt[3]{\frac{\sin^2(x)}{\cos^2(x)}}} dx$$

$$= \int_{\frac{\pi}{6}}^{\frac{\pi}{4}} \tan^{\frac{-2}{3}}(x)\left(1 + \tan^2(x)\right) dx \qquad (1)$$

Now, we should change the variable of the integral as follows.

$$\tan(x) \triangleq u \stackrel{\frac{d}{dx}}{\Rightarrow} \left(1 + \tan^2(x)\right) dx = du \qquad (2)$$

Solving (1) and (2):

$$\int_{u_1}^{u_2} u^{-\frac{2}{3}} du = 3u^{\frac{1}{3}}\Big|_{u_1}^{u_2} = \left(3\ \tan^{\frac{1}{3}}(x)\right)\Big|_0^{\frac{\pi}{4}} = 3(1 - 0) = 3$$

Choice (3) is the answer.

8.54. From the list of integral of trigonometric functions, we know that:

$$\int \frac{1}{u} du = \ln|u| + c$$

Based on the information given in the problem, we have:

$$f(1) = 0 \qquad (1)$$

$$\frac{d}{dx}\left(f(x^2)\right) = \frac{6}{x} \qquad (2)$$

The problem can be solved as follows.

$$\frac{d}{dx}\left(f(x^2)\right) = 2xf'(x^2) \qquad (3)$$

$$\stackrel{(2),(3)}{\Longrightarrow} \frac{6}{x} = 2xf'(x^2) \Rightarrow f'(x^2) = \frac{3}{x^2} \qquad (4)$$

By changing the variable of the integral, we have:

$$x^2 \triangleq t \qquad (5)$$

$$\xrightarrow{(4),(5)} f'(t) = \frac{3}{t} \xrightarrow{\int dt} f(t) = 3\ln|t| + c \tag{6}$$

$$\xrightarrow{(1),(6)} 0 = 3 \times 0 + c \Rightarrow c = 0 \Rightarrow f(t) = 3\ln|t|$$

$$\Rightarrow f(e) = 3\ln(e) = 3 \times 1 = 3$$

Choice (3) is the answer.

8.55. From the list of integral of trigonometric functions, we know that:

$$\int x^n dx = \frac{1}{n+1} x^{n+1} + c$$

From trigonometry, we know that:

$$1 + \cos(2x) = 2\cos^2(x)$$

Moreover, based on the information given in the problem, we have:

$$f(1) = 1 \tag{1}$$

$$f'\left(\cos^2(x)\right) = \cos(2x) \tag{2}$$

The problem can be solved as follows.

$$f'\left(\cos^2(x)\right) = \cos(2x) = 2\cos^2(x) - 1 \tag{3}$$

By changing the variable of the integral, we have:

$$\cos^2(x) \triangleq t \tag{4}$$

$$\xrightarrow{(3),(4)} f'(t) = 2t - 1 \xrightarrow{\int dt} f(t) = t^2 - t + c \tag{5}$$

$$\xrightarrow{(1),(5)} 1 = 1 - 1 + c \Rightarrow c = 1 \Rightarrow f(t) = t^2 - t + 1$$

$$\Rightarrow f(-1) = (-1)^2 - (-1) + 1 = 3$$

Choice (3) is the answer.

8.56. From the list of integral of trigonometric functions, we know that:

$$\int x^n dx = \frac{1}{n+1} x^{n+1} + c$$

From trigonometry, we know that:

$$\sin(2x) = 2\sin(x)\cos(x)$$

The problem can be solved as follows.

$$\int \frac{\cos{(2x)}}{\sin^2{(x)}\cos^2{(x)}}dx = \int \frac{\cos{(2x)}}{\left(\frac{1}{2}\sin{(2x)}\right)^2}dx = \int \frac{\cos{(2x)}}{\frac{1}{4}\sin^2{(2x)}}dx = \int 4\cos{(2x)}\left(\sin{(2x)}\right)^{-2}dx$$

Now, we need to change the variable of the integral as follows.

$$\sin{(2x)} \triangleq u \xrightarrow{\frac{d}{dx}} 2\cos{(2x)}dx = du$$

$$\Rightarrow \int 2u^{-2}du = -\frac{2}{u}+c = \frac{-2}{\sin{(2x)}}+c$$

Choice (1) is the answer.

8.57. From integration by parts (partial integration), we know that:

$$\int \ln{(x)}dx = x\ln{|x|} - x$$

or, in general:

$$\int u(x)dv = u(x)v(x) - \int v(x)du$$

In addition, from the list of integral of trigonometric functions, we know that:

$$\int \frac{1}{u}du = \ln{|u|}+c$$

$$\int x^n dx = \frac{1}{n+1}x^{n+1}+c$$

The problem can be solved as follows.

$$\int_1^e (2x + \ln{(x)})dx = \int_1^e 2x\,dx + \int_1^e \ln{(x)}dx = \left(x^2\right)\Big|_1^e + \left(x\ln{|x|}-x\right)\Big|_1^e$$

$$= e^2 - 1 + (e-e) - (0-1) = e^2$$

Choice (1) is the answer.

8.58. From the list of integral of trigonometric functions, we know that:

$$\int x^n dx = \frac{1}{n+1}x^{n+1}+c$$

The problem can be solved as follows.

$$\int \sin{(2x)}\left(2 + \cos^2{(x)}\right)^{50}dx \tag{1}$$

We should change the variable of the integral as follows.

$$2 + \cos^2(x) \triangleq u \implies -2\cos(x)\sin(x)dx = du \implies -\sin(2x) = du \tag{2}$$

Solving (1) and (2):

$$-\int u^{50}du = -\frac{u^{51}}{51} + c = -\frac{1}{51}\left(2 + \cos^2(x)\right)^{51} + c$$

Choice (2) is the answer.

8.59. From the list of integral of trigonometric functions, we know that:

$$\int u^n du = \frac{1}{n+1}u^{n+1} + c$$

The problem can be solved as follows.

$$\int_1^e \frac{\ln(x)}{x}dx = \int_1^e \ln(x)\left(\frac{1}{x}\right)dx \tag{1}$$

Now, we should change the variable of the integral as follows.

$$\ln(x) \triangleq u \implies \frac{1}{x}dx = du \tag{2}$$

Solving (1) and (2):

$$\int_{u_1}^{u_2} u\,du = \frac{1}{2}u^2\Big|_{u_1}^{u_2} = \frac{1}{2}\left(\ln(x)\right)^2\Big|_1^e = \frac{1}{2} - 0 = \frac{1}{2}$$

Choice (2) is the answer.

8.60. From the list of integral of trigonometric functions, we know that:

$$\int u^n du = \frac{1}{n+1}u^{n+1} + c$$

The problem can be solved as follows.

$$y' = -\frac{2x+2}{4y+1} \implies 4yy' + y' = -2x - 2 \xrightarrow{\int dx} 2y^2 + y = -x^2 - 2x + c \tag{1}$$

$$\xrightarrow{(x,y) = (0,1)} 2 + 1 = 0 + c \implies c = 3 \tag{2}$$

Solving (1) and (2):

$$2y^2 + y = -x^2 - 2x + 3 \implies x^2 + 2y^2 = -y - 2x + 3$$

Choice (4) is the answer.

8.61. From the list of integral of trigonometric functions, we know that:

$$\int u^n du = \frac{1}{n+1} u^{n+1} + c$$

Moreover, from trigonometry, we know that:

$$\cot(x) = \frac{\cos(x)}{\sin(x)}$$

$$1 - \cos(2x) = 2\sin^2(x)$$

The problem can be solved as follows.

$$\int_{\frac{\pi}{6}}^{\frac{\pi}{2}} \frac{\cot(x)}{\sqrt{1-\cos(2x)}} dx = \int_{\frac{\pi}{6}}^{\frac{\pi}{2}} \frac{\cos(x)}{\sqrt{2\sin^2(x)}\sin(x)} dx = \int_{\frac{\pi}{6}}^{\frac{\pi}{2}} \frac{\cos(x)}{\sqrt{2}|\sin(x)|\sin(x)} dx$$

$$= \frac{\sqrt{2}}{2} \int_{\frac{\pi}{6}}^{\frac{\pi}{2}} (\sin(x))^{-2} \cos(x) dx \qquad (1)$$

Now, we should change the variable of the integral as follows.

$$\sin(x) \triangleq u \Rightarrow \cos(x)dx = du \qquad (2)$$

Solving (1) and (2):

$$\frac{\sqrt{2}}{2} \int_{u_1}^{u_2} u^{-2} du = -\frac{\sqrt{2}}{2} u^{-1} \Big|_{u_1}^{u_2} = \left(-\frac{\sqrt{2}}{2} \times \frac{1}{\sin(x)}\right)\Big|_{\frac{\pi}{6}}^{\frac{\pi}{2}} = -\frac{\sqrt{2}}{2}(1-2) = \frac{\sqrt{2}}{2}$$

Choice (2) is the answer.

8.62. From the list of integral of trigonometric functions, we know that:

$$\int u^n du = \frac{1}{n+1} u^{n+1} + c$$

The problem can be solved as follows.

$$\int_3^6 \frac{x+2}{\sqrt{x-2}} dx = \int_3^6 \frac{x-2+4}{\sqrt{x-2}} dx = \int_3^6 \left((x-2)^{\frac{1}{2}} + 4(x-2)^{-\frac{1}{2}}\right) dx$$

$$= \left(\frac{2}{3}(x-2)^{\frac{3}{2}} + 4 \times 2(x-2)^{\frac{1}{2}}\right)\Big|_3^6 = \left(\frac{2}{3}(4)^{\frac{3}{2}} + 4 \times 2(4)^{\frac{1}{2}}\right) - \left(\frac{2}{3}(1)^{\frac{3}{2}} + 4 \times 2(1)^{\frac{1}{2}}\right)$$

$$= \left(\frac{16}{3} + 16\right) - \left(\frac{2}{3} + 8\right) = \frac{14}{3} + 8 = \frac{38}{3}$$

Choice (2) is the answer.

8.63. In addition, from the list of integral of trigonometric functions, we know that:

$$\int u^n du = \frac{1}{n+1} u^{n+1} + c$$

The problem can be solved as follows.

$$\int_1^4 \frac{\sqrt{1+\sqrt{x}}}{\sqrt{x}} dx = 2 \int_1^4 \left(1 + \sqrt{x}\right)^{\frac{1}{2}} \times \frac{1}{2\sqrt{x}} dx \tag{1}$$

Now, we should change the variable of the integral as follows.

$$1 + \sqrt{x} \triangleq u \Longrightarrow \frac{1}{2\sqrt{x}} dx = du \tag{2}$$

Solving (1) and (2):

$$2 \int_{u_1}^{u_2} u^{\frac{1}{2}} du = \left(2 \times \frac{2}{3} u^{\frac{3}{2}}\right)\Big|_{u_1}^{u_2} = \left(\frac{4}{3}\left(1+\sqrt{x}\right)^{\frac{3}{2}}\right)\Big|_1^4 = \frac{4}{3}\left(3\sqrt{3} - 2\sqrt{2}\right) = 4\left(\sqrt{3} - \frac{2\sqrt{2}}{3}\right)$$

Choice (4) is the answer.

8.64. In addition, from the list of integral of trigonometric functions, we know that:

$$\int u^n du = \frac{1}{n+1} u^{n+1} + c$$

From trigonometry, we know that:

$$\cot(x) = \frac{\cos(x)}{\sin(x)}$$

The problem can be solved as follows.

$$\int \cot(x)\sqrt{\sin(x)} dx = \int \frac{\cos(x)}{\sin(x)} \left(\sin(x)\right)^{\frac{1}{2}} dx = \int \left(\sin(x)\right)^{\frac{-1}{2}} \cos(x) dx \tag{1}$$

Now, we should change the variable of the integral as follows.

$$\sin(x) \triangleq u \Longrightarrow \cos(x) dx = du \tag{2}$$

Solving (1) and (2):

$$\int u^{\frac{-1}{2}} du = 2u^{\frac{1}{2}} + c = 2\sqrt{\sin(x)} + c$$

Choice (2) is the answer.

8.65. From the list of integral of trigonometric functions, we know that:

$$\int x^n du = \frac{1}{n+1} x^{n+1} + c$$

The volume resulted from the rotation of a surface area around y-axis, enclosed between the curve of $f(x)$ and x-axis, is calculated as follows:

$$V = \pi \int_{y_1}^{y_2} x^2 dy$$

Since x-axis is the boundary, $y_1 = 0$. Another boundary for y can be determined as follows:

$$x = 0 \Longrightarrow y_2 = 1 - \frac{1}{4} \times 0 = 1$$

Therefore,

$$V = \pi \int_0^1 x^2 dy = \pi \int_0^1 (4 - 4y) dy = \pi \left(4y - 2y^2\right)\Big|_0^1 = \pi(4 - 2) = 2\pi$$

Choice (2) is the answer.

8.66. From trigonometry, we know that:

$$\sec(x) = \frac{1}{\cos(x)}$$

$$\tan(x) = \frac{\sin(x)}{\cos(x)}$$

The problem can be solved as follows.

$$\int_0^{\frac{\pi}{3}} \sec(x) \tan(x) dx = \int_0^{\frac{\pi}{3}} \frac{1}{\cos(x)} \frac{\sin(x)}{\cos(x)} dx = \int_0^{\frac{\pi}{3}} \frac{\sin(x)}{\cos^2(x)} dx \qquad (1)$$

Now, we should change the variable of the integral as follows.

$$\cos(x) \triangleq u \Longrightarrow -\sin(x) dx = du \qquad (2)$$

Solving (1) and (2):

$$-\int_{u_1}^{u_2} \frac{1}{u^2} du = \frac{1}{u}\Big|_{u_1}^{u_2} = \frac{1}{\cos(x)}\Big|_0^{\frac{\pi}{3}} = \frac{1}{\frac{1}{2}} - \frac{1}{1} = 1$$

Choice (1) is the answer.

8.67. From trigonometry, we know that:

$$\csc(x) = \frac{1}{\sin(x)}$$

$$\cot(x) = \frac{\cos(x)}{\sin(x)}$$

The problem can be solved as follows.

$$\int_{\frac{\pi}{6}}^{\frac{\pi}{4}} \csc(x) \cot(x) dx = \int_{\frac{\pi}{6}}^{\frac{\pi}{4}} \frac{1}{\sin(x)} \frac{\cos(x)}{\sin(x)} dx = \int_{\frac{\pi}{3}}^{\frac{\pi}{4}} \frac{\cos(x)}{\sin^2(x)} dx \tag{1}$$

Now, we should change the variable of the integral as follows.

$$\sin(x) \triangleq u \Rightarrow \cos(x) dx = du \tag{2}$$

Solving (1) and (2):

$$\int_{u_1}^{u_2} \frac{1}{u^2} du = -\frac{1}{u} \Big|_{u_1}^{u_2} = -\frac{1}{\sin(x)} \Big|_{\frac{\pi}{6}}^{\frac{\pi}{4}} = \left(\frac{1}{\frac{\sqrt{2}}{2}} - \frac{1}{\frac{1}{2}} \right) = -\left(\sqrt{2} - 2 \right) = 2 - \sqrt{2}$$

Choice (2) is the answer.

8.68. From trigonometry, we know that:

$$1 + \cot^2(x) = \frac{1}{\sin^2(x)}$$

$$1 + \tan^2(x) = \frac{1}{\cos^2(x)}$$

$$\tan(x) \cot(x) = 1$$

Moreover, from the list of integral of trigonometric functions, we know that:

$$\int \left(1 + \tan^2(x) \right) dx = \tan(x) + c$$

$$\int \left(1 + \cot^2(x) \right) dx = -\cot(x) + c$$

The problem can be solved as follows.

$$\int_{\frac{\pi}{6}}^{\frac{\pi}{4}} \frac{1}{\sin^2(x) \cos^2(x)} dx = \int_{\frac{\pi}{6}}^{\frac{\pi}{4}} \left(1 + \cot^2(x) \right) \left(1 + \tan^2(x) \right) dx$$

$$= \int_{\frac{\pi}{6}}^{\frac{\pi}{4}} \left(1 + \tan^2(x) + \cot^2(x) + \cot^2(x) \tan^2(x) \right) dx = \int_{\frac{\pi}{6}}^{\frac{\pi}{4}} \left(1 + \tan^2(x) + \cot^2(x) + 1 \right) dx$$

$$\int_{\frac{\pi}{6}}^{\frac{\pi}{4}} \left(1 + \tan^2(x) \right) dx + \int_{\frac{\pi}{6}}^{\frac{\pi}{4}} \left(1 + \cot^2(x) \right) dx = \left(\tan(x) - \cot(x) \right) \Big|_{\frac{\pi}{6}}^{\frac{\pi}{4}}$$

$$= (1 - 1) - \left(\frac{\sqrt{3}}{3} - \sqrt{3} \right) = \frac{2\sqrt{3}}{3}$$

Choice (2) is the answer.

8.69. From the list of integral of trigonometric functions, we know that:

$$\int x^n dx = \frac{1}{n+1} x^{n+1} + c$$

The problem can be solved as follows.

$$\int (\tan(x) - \cot(x))(\tan(x) + \cot(x))^5 dx$$

$$= \int (\tan(x) - \cot(x))(\tan(x) + \cot(x))(\tan(x) + \cot(x))^4 dx$$

$$= \int (\tan^2(x) - \cot^2(x))(\tan(x) + \cot(x))^4 dx$$

$$= \int (1 + \tan^2(x) - (1 + \cot^2(x)))(\tan(x) + \cot(x))^4 dx \qquad (1)$$

Now, we need to change the variable of the integral as follows.

$$\tan(x) + \cot(x) \triangleq u \xrightarrow{\frac{d}{dx}} (1 + \tan^2(x) - (1 + \cot^2(x))) dx = du \qquad (2)$$

Solving (1) and (2):

$$\int u^4 du = \frac{u^5}{5} + c = \frac{1}{5}(\tan(x) + \cot(x))^5 + c$$

Choice (2) is the answer.

8.70. The volume resulted from the rotation of a surface area around x-axis, enclosed between the curve of $f(x)$ and x-axis, is calculated as follows:

$$V = \pi \int_{x_1}^{x_2} (f(x))^2 dx$$

Moreover, from trigonometry, we know that:

$$1 + \cot^2(x) = \frac{1}{\sin^2(x)}$$

In addition, from the list of integral of trigonometric functions, we know that:

$$\int u^n du = \frac{1}{n+1} u^{n+1} + c$$

$$\int (1 + \cot^2(x)) dx = -\cot(x)$$

Therefore,

$$V = \pi \int_{\frac{\pi}{4}}^{\frac{\pi}{2}} \frac{1}{\sin^4(x)} dx = \pi \int_{\frac{\pi}{4}}^{\frac{\pi}{2}} (1 + \cot^2(x))(1 + \cot^2(x)) dx \qquad (1)$$

Now, we should change the variable of the integral as follows.

$$\cot(x) \triangleq u \Rightarrow -\left(1 + \cot^2(x)\right)dx = du \tag{2}$$

Solving (1) and (2):

$$V = -\pi \int_{u_1}^{u_2} \left(1 + u^2\right)du = -\pi\left(u + \frac{1}{3}u^3\right)\Big|_{u_1}^{u_2} = -\pi\left(\cot(x) + \frac{1}{3}\cot^3(x)\right)\Big|_{\frac{\pi}{4}}^{\frac{\pi}{2}}$$

$$V = -\pi\left((0+0) - \left(1 + \frac{1}{3}\right)\right) = \frac{4}{3}\pi$$

Choice (4) is the answer.

8.71. From trigonometry, we know that:

$$1 + \cos(2x) = 2\cos^2(x) \tag{1}$$

$$\sin^2(x) + \cos^2(x) = 1 \tag{2}$$

The problem can be solved as follows.

$$I = \int \sin(x)\cos(x)dx \tag{3}$$

Now, we should change the variable of the integral as follows.

$$\sin(x) \triangleq u \Rightarrow \cos(x)dx = du \tag{4}$$

Solving (3) and (4):

$$I = \int u\,du = \frac{1}{2}u^2 + c = \frac{1}{2}\sin^2(x) + c \tag{5}$$

Solving (2) and (5):

$$I = \frac{1}{2}\left(1 - \cos^2(x)\right) + c = -\frac{1}{2}\cos^2(x) + \frac{1}{2} + c = -\frac{1}{2}\cos^2(x) + c' \tag{6}$$

Solving (1) and (6):

$$I = -\frac{1}{2}\left(\frac{1}{2} + \frac{1}{2}\cos(2x)\right) + c' = -\frac{1}{4}\cos(2x) + c' - \frac{1}{4} = -\frac{1}{4}\cos(2x) + c'' \tag{7}$$

From (5), (6), and (7), choice (2) is the answer.

Reference

1. Rahmani-Andebili, M. (2020). Precalculus: Practice problems, methods, and solutions, Springer Nature, 2020.

Index

A
Absolute extrema, 88, 90, 101, 104
Alternate notation, 10, 11, 35, 36
Applications of derivatives, 85, 88, 89, 95, 100, 102
Applications of integrals, 111, 113, 115, 117, 118, 121, 129, 133, 134, 136, 139, 141, 146
Application, Taylor series in limits, 64–66, 77–82
Area between curves, 111–113, 117, 123, 129–131, 138, 151
Average value, 109, 115, 122, 127, 135, 148

C
Chain rule, 83–87, 91–94, 96, 99
Co-function formulas, 4, 16, 25, 47
Critical points, 88, 101

D
Definite integral, 107, 108, 110, 112, 114, 116, 125–127, 130, 134, 137
Definition of derivative, 89, 103
Degrees to radians formulas, 1, 4, 9, 17, 19, 24, 33, 49
Derivatives of exponential, 83, 88, 91, 101
Derivatives of hyperbolic functions, 86, 90, 96, 104
Derivatives of inverse trigonometric functions, 85, 87, 94, 99
Derivatives of trigonometric functions, 85, 94–96
Differentiation formulas, 84, 87, 88, 93, 99, 101
Double angle formulas, 14, 41, 42

E
Even and odd formulas, 5, 16, 26, 27, 48

H
Half angle formulas, 4, 8, 24, 31
Higher-order derivatives, 83, 92

I
Implicit differentiation, 84, 86, 88, 89, 92, 93, 97, 98, 100, 102
Indefinite integral, 107, 108, 111, 118, 123, 125, 130, 141, 151
Integrals involving quadratics, 111, 116, 128, 137
Integrals involving roots, 108–110, 114, 121, 126–128, 133, 147
Integrals involving trigonometric functions, 108, 110, 111, 115, 120, 126, 128, 135, 136, 144
Integration by parts, 109, 112, 114, 121–123, 126, 130, 133, 134, 146, 148, 150
Integration techniques, 112, 113, 118, 120, 131, 132, 140, 141, 145
Integration using partial fractions, 117, 119, 138, 143
Inverse properties, 5, 6, 9, 11, 12, 14, 27, 28, 32, 37, 38, 43
Inverse trigonometric functions, 4, 5, 7, 10, 14, 26, 30, 33, 34, 43

L
Least common multiple (LCM), 21
Limits and continuity, 64, 79
Limits at infinity, 54, 55, 57, 58, 60, 61, 67, 68, 70, 71, 73, 74
Limits by direct substitution, 53, 56, 58, 67, 69, 71
Limits by factoring, 54, 55, 60–63, 68, 73–75, 77
Limits by L'Hopital's rule, 58, 59, 71, 72
Limits by rationalization, 57, 59, 61, 63, 70, 72, 73, 76
Limits involving Euler's number, 58, 60, 62–65, 71, 73–76, 78, 80
Limits of absolute value functions, 53–55, 58, 67, 68, 72

M
Minimum and maximum values, 86, 89, 90, 97, 102, 103

P
Period, 2, 3, 21, 23
Periodic formulas, 2, 4, 20, 21, 25
Product rule, 84, 93
Product to sum formulas, 3, 11, 17, 18, 24, 36, 50, 51
Pythagorean identities, 15, 17, 44, 46, 49, 50

Q
Quotient rule, 87, 98

R
Range, 2, 7, 12, 13, 16, 20, 30, 39, 40, 48
Rates of change, 89
Reciprocal identities, 14, 15, 17, 42, 44, 49

S
Sine and cosine identities, 1, 16, 20, 47
Substitution rule for integrals, 113, 116, 123, 132, 137, 149
Sum and difference to product formulas, 7, 18, 30, 51

T
Tangent and cotangent identities, 1, 19

© Springer Nature Switzerland AG 2021
M. Rahmani-Andebili, *Calculus*, https://doi.org/10.1007/978-3-030-64980-7

Printed in the United States
by Baker & Taylor Publisher Services